The Natural Features of Mount Fuji

Volume 2
Satoyama Environments in Crisis

This photo depicts the annual burning of grasslands that has been taking place for approximately 70 years. Although they developed this practice to support their livelihoods, locals have inadvertently preserved many endangered plants and animals through their maintenance of the grasslands. Recently, however, the *satoyama* environments maintained by human intervention have been in decline. All of us should consider how to protect the precious natural environment that still exists around Mt. Fuji.

Nonprofit Organization
Mount Fuji Nature Conservation Center

Contents

1. The three types of nature. 3
2. Transformed *satoyama* nature 4
3. *Satoyama* environments around Mt. Fuji 6
4. The three elements of *satoyama* nature (woodlands, grasslands, and waterfronts) 7
5. Current status of plants and animals living in the *satoyama* environments of Mt. Fuji ... 8
    I. Current status of plants and animals living in *satoyama* grassland environments 8
        A. Why are the *satoyama* grassland environments important? 8
        B. *Satoyama* grasslands maintained by human activities 10
        C. Typical threatened grassland plants 13
        D. Endangered grassland insects 20
        E. Major threatened grassland birds 28
        F. Threatened grassland mammals 34
    II. Current status of plants and animals living in *satoyama* woodland environments 36
        A. Why are the *satoyama* woodland environments important? 36
        B. Typical threatened woodland plants 38
        C. Representative threatened insects 42
        D. Representative threatened birds 44
        E. Threatened mammals 50
        F. Rare amphibians 52
        G. Why are *satoyama* lava flows important? 54
        H. Animals uniquely adapted to living in lava fields 58
    III. Current status of plants and animals living in *satoyama* waterfront environments ... 62
        A. Why are *satoyama* waterfront environments important? 62
        B. Representative wetland plants that are threatened 64
        C. Rare Amphibians · Reptiles 70
6. Thinking about the future of Mt. Fuji's *satoyama* 72
    I. The problem of invasive plants and animals 72
    II. Things to consider when experiencing the nature of Mt. Fuji 74
        A. Mt. Fuji as a national park 74
        B. Nature's place in the Mt. Fuji World Cultural Heritage Site 76
    III. Species thought to be extinct on the northern side of Mt. Fuji 77
7. Mount Fuji Nature Conservation Center (NPO) 78

References & Index 79

1. Broadly speaking, there are three types of nature.

Many people think of wilderness when they hear the term "nature," but we can divide nature into three main types. How we ought to interact with nature depends on what kind of nature we are talking about.

A. Wilderness

Nature undisturbed by humans for long stretches of time.
→ This type of nature can be preserved without human intervention (conservation narrowly defined).

C. Park-like nature (urban nature)

Nature that consists of trees, grass, and other plants cultivated by humans.
→ This form of nature depends on regular horticultural management.

B. *Satoyama* (upland countryside) nature

This type of nature, sometimes called "secondary nature," is maintained through periodic human intervention.
→ Human activities such as mowing, burning, and coppicing work to conserve this type of nature.

2. Transformed *satoyama* nature

Satoyama (upland countryside) is a term now used worldwide—for example, a "*satoyama* initiative" was adopted at the COP10 biodiversity treaty conference held in Nagoya 2010—but it is understood in various ways.

Here we use *satoyama* to refer to "an area close to human settlements where woodlands are regularly coppiced for building supplies as well as firewood and charcoal, and where grasslands are maintained through periodic mowing and burning in order to provide fertilizer. In addition to the woodlands and grasslands maintained by humans to generate resources for agriculture and daily living, the *satoyama* environment consists of paddy fields, orchards, and villages that depend on these activities and resources" (Shigematsu et al., 2010).

Moreover, "most *satoyama* forests, including red pine forests, consist of copses and woodlands used for firewood and agricultural activities. They are called 'secondary forests' because of their transitional state. People refer to grasslands with terms such as *kayaba* (place that has thatch for roofing) or *magusaba* (place to rear farm animals)" (Shigematsu et al., 2010).

Until the late 1950s, *satoyama* environments were integral to farmers' lives. To support themselves, farmers periodically coppiced woodlands as well as mowed grasslands to raise livestock and thatch roofs. They also maintained paddies and fields to grow rice and vegetables. Various plants and animals that are now endangered coexisted with people in these environments.

Because of the adoption of fuel oil and chemical fertilizers, however, people no longer make charcoal, mow grass, or coppice woodlands. People have also abandoned cultivated fields, leaving many of them

fallow. Moreover, because they have largely stopped raising livestock and using thatch for the roofs of houses, they no longer mow grasslands, which are now turning into forest. Finally, the decline in forestry has led to the abandonment of many plantation forests.

Because of these changes to *satoyama* environments, plants and animals that were once commonly found around the houses and fields of farmers—such as the balloon flower, killifish, and black-spotted pond frog—have become threatened species that are vanishing all around us. Mt. Fuji and the surrounding region are no exception. We should cherish the valuable *satoyama* environments that are now shrinking, and think about how to coexist with the precious animals and plants that live there.

Satoyama before the late 1950s

natural forest | maintained coppice | grassland | cultivated land | creek | paddy field | garden | farmer's house

Economic change → **Lifestyle change**

Satoyama now

natural forest | abandoned plantation forest | abandoned coppice | abandoned cultivated land | paddy field | park | pond | apartment building

3. Most of the nature around Mt. Fuji consists of *satoyama* environments.

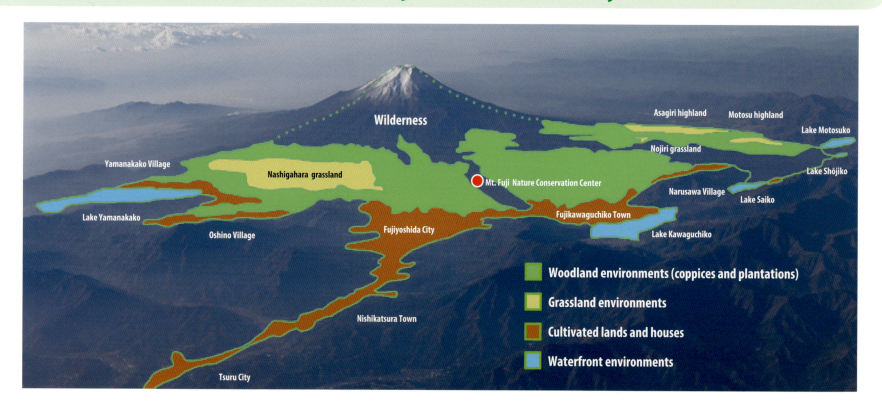

Most people think of relatively untouched wilderness when they hear "nature of Mt. Fuji," but as this map indicates, most of the nature around Fuji actually consists of *satoyama* environments. These include grasslands such as the Nashigahara, Motosukōgen, and Nojirisōgen, as well as woodland environments comprising both coppice and large plantation forests. They also include waterfront environments like the shores of Mt. Fuji's rivers and five lakes.

Therefore, when thinking about the "nature of Mt. Fuji," we must first and foremost consider how we should preserve these *satoyama* environments.

4. The three elements of *satoyama* nature (woodlands, grasslands, and waterfronts)

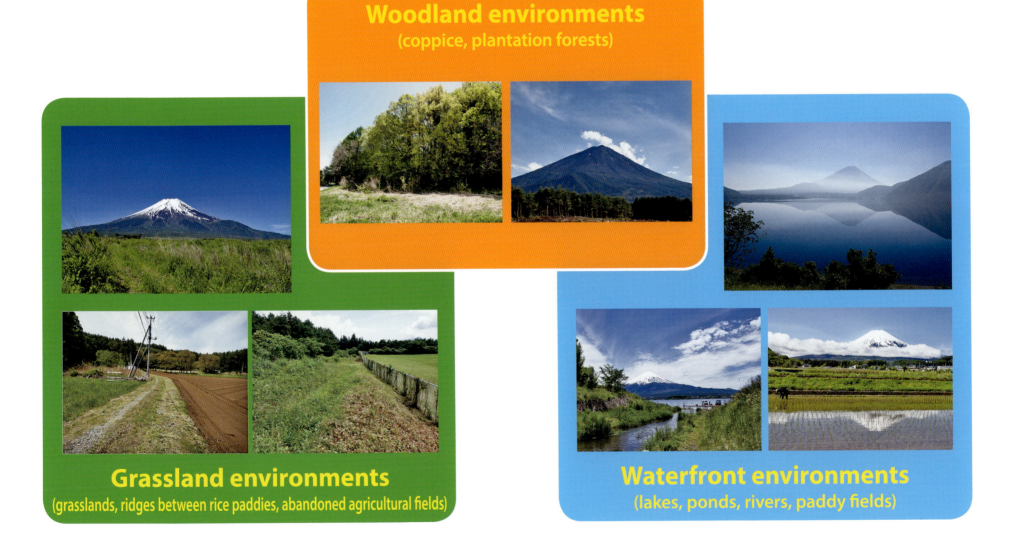

5. Current status of plants and animals living in the *satoyama* environments of Mt. Fuji

I. Current status of plants and animals living in *satoyama* grassland environments

A. Why are the *satoyama* grassland environments important?

A Japanese beech forest, one of the "100 forests" of Yamanashi Prefecture, located near the second station of the Shōji climbing trail

The northern foot of Mt. Fuji is situated in the middle of the Japanese archipelago, which is in the temperate monsoon climate zone and therefore has distinct seasons. In terms of vegetation, it is located in the deciduous broadleaf forest zone, characterized by forests of Japanese beech and Mongolian oak (Miyawaki & Sugawara, 1992). So as the years pass, bare land turns into grassland (consisting of annual and perennial plants as well as sparse groves of trees), then sun-tolerant woodland, and finally shade-tolerant woodland (climax forest). This process of change in vegetation is called plant succession, with the change from bare land called primary succession (Ishii et al., 1993; The Ecological Society of Japan, 2012).

A schematic diagram of plant succession at the northern foot of Mt. Fuji appears on the next page. It differs from other ones in that it illustrates standard soil development together with the main tree species that live at the northern foot of Mt. Fuji.

Here it is worth noting that *satoyama* environments, including both fields and coppice, have been maintained in Japan since the Yayoi period (ca. 600 BCE-300 CE) for agricultural purposes. As noted on pages 4 and 5, this was the natural result of farmers working to support their everyday lives. Among other things, they maintained grasslands through burning in order to create thatch for roofs and meadows for agriculture. However, burning now occurs in only limited areas, and there has been a decline in grassland environments bordering agricultural lands, including the ridges between paddies and fields. Consequently, the remaining *satoyama* grasslands at the northern foot of Mt. Fuji are crucial to the endangered plants and animals that live there. Reflecting a trend seen throughout Japan, many of these species are registered in the Red Lists (lists of endangered and threatened species) maintained by the Ministry of the Environment as well as Yamanashi and Shizuoka prefectures.

At this point we will introduce the current status of these endangered plants and animals, and we encourage everyone to consider how best to preserve them.

Meadows of Japanese pampas grass

Scattered groves of Japanese oak

Meadows and coppice with groves of sun-tolerant trees

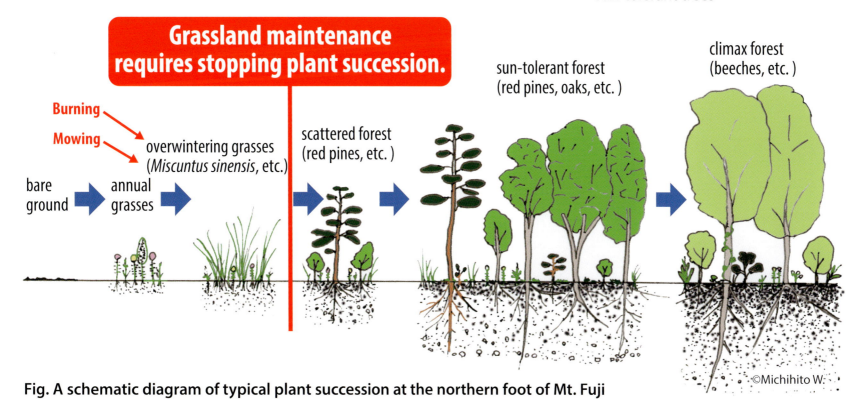

Fig. A schematic diagram of typical plant succession at the northern foot of Mt. Fuji

I. Current status of *satoyama* grassland plants and animals

B. *Satoyama* grasslands maintained by human activities

① Grassland maintained by yearly burning (Nashigahara)

Burning in April

Two months after burning

Four months after burning

② Crucial grassland maintained after cutting

About one year after cutting

A few years after planting

About ten years after planting

③ Grassland maintained by yearly mowing (Motosukōgen)

The firebreak zone that appears on the left-hand side of this photo is mowed every November. The cut grass is piled up on both sides of the zone.

April 17, 2012   June 1, 2012   August 25, 2012

The trail that appears in the middle of the bottom photo is mowed every November as well.

April 17, 2012   June 1, 2012   August 25, 2012

In the following pages, we employ the Red Lists published by Japan's Ministry of the Environment (J) in 2017 and by Yamanashi Prefecture (Y) in 2018 and Shizuoka Prefecture (S) in 2017 to categorize species according to the degree they are endangered. The categories are as follows: CR (Critically Endangered), EN (Endangered), VU (Vulnerable), NT (Nearly Threatened). We indicate if data is deficient (DD) or something is especially noteworthy (N), and also if a rare species is as yet unpublished in the Red Lists (Un) but will likely be included in future editions.

An outline of the rankings determined by the Ministry of the Environment of Japan appears below. The ministry's Red List covers species nationwide, while those published by Yamanashi and Shizuoka prefectures focus on species found in those regions.

Grassland

Red List Categories

according to IUCN web site

Category I (CR+EN)	Category IA: Critically Endangered (CR)	Extremely high risk of extinction in the wild.
	Category IB: Endangered (EN)	Very high risk of extinction in the wild.
Category II: Vulnerable (VU)		High risk of extinction in the wild.
Nearly Threatened (NT)		Close to qualifying as threatened in the near future.
Data Deficient (DD)		Inadequate information to make a direct, or indirect, assessment of extinction risk based on distribution and/or population status.

These species are at high risk if removed from their habitats, so the following photos depict these endangered animals and plants in their natural settings. By showing where these species live, the photos also underscore the importance of the *satoyama* environment around Mt. Fuji and encourage us to consider how best to conserve these threatened plants and animals.

I. Current status of the plants and animals living in grassland environments

C. Typical threatened grassland plants

The extensive grasslands found at the northern foot of Mt. Fuji are home to over 250 grassland species.

Okinagusa (*Pulsatilla cernua*) (VU(J)/EN(Y/S))

In the past, this plant was widespread in areas with soil mixed together with gravel, but now it is rare. Today we can see this plant in part of the grasslands at the northern foot of Mt. Fuji.

Kikyō (Balloon flower) (*Platycodon grandiflorus*) (VU(J)/ NT(Y)/VU(S))

One of the "seven herbs of autumn" (**akinonanakusa**), this threatened species is in decline throughout Japan. We used to see this flower at many places in Yamanashi and Shizuoka prefectures, but now it is rare to find them. Fortunately it is still easy to find in parts of the grasslands at the northern foot of Mt. Fuji.

This plant is called **okinagusa** (old man grass) because the white hair on its fruit looks like the hair of an old man.

Okinagusa flower

Kikyō bud and flower

Grassland

Kōrinka (*Tephroseris flammea*) (VU(J)/ NT(Y)/EN(S))

This plant has become extremely rare nationwide, and we can now see it only in the extensive grasslands at the northern foot of Mt. Fuji. In certain places in the grasslands we can find a significant number of these flowers, but there are also many spots where they are rarely seen.

Senburi (*Swertia japonica*) (NT(Y))

This species used to be widespread in the grasslands of Yamanashi Prefecture, but its numbers have recently been in decline. We can see clusters of them only in specific places at the northern foot of Mt. Fuji.

It is called **kōrinka** (red circle flower) because its vivid reddish-orange petals are arranged in a circle.

The long-blooming **kōrinka** flower is an important source of nectar for many insects.

Senburi flower
In the past this plant was gathered as a medicinal herb.

Murasaki-senburi (*Swertia pseudochinensis*) (NT(J)/EN (Y)/VU(S))

This plant is found in the Kantō region and western Japan, but it is now in decline. Its habitat is limited in Yamanashi Prefecture, and it can be seen in only certain places on and around the northern foot of Mt. Fuji.

Suzusaiko (*Vincetoxicum pycnostelma*) (NT(J)/VU(Y)/NT(S))

A member of the subfamily *Asclepiadoideae* ※ , this plants blooms in the morning during the summer. Fortunately it can still be found in relative abundance in the grasslands at the northern foot of Mt. Fuji.

※ According to the APG (Angiosperm Phylogeny Group) plant classification system

Murasaki-senburi bud and flower

Suzusaiko flower

The bud of the **suzusaiko** is shaped like a round bell (*suzu* means bell).

Grassland

Bāsobu (Himetsuruninjin) (*Codonopsis ussuriensis*) (VU(J)/EN(Y)/VU(S))

A member of the *Campanulacea* family, this plant lives in grasslands throughout Japan, but its numbers are declining. In Yamanashi and Shizuoka prefectures, its numbers are shrinking along with its habitat. This plant blooms from August to October. It resembles the closely related species **jisobu**, but its leaves have no hair and its flower is a bit larger.

Bāsobu bud and flower

Mizuchidori (*Platanthera hologlottis*) (VU(Y/S))

Distributed throughout Japan, it lives in the montane grasslands of Yamanashi and Shizuoka prefectures. Its white flower blooms from June to July, and it is called jakōchidori because it smells like fragrant incense (*jakō*).

Mizuchidori flower

Hinanokinchaku (*Polygala tatarinowii*) (EN(J)/CR(Y)/EN(S))

An annual plant in the *Polygalaceae* family. The habitat and numbers of these plants are extremely rare in Japan, and we rarely see this plant at the northern foot of Mt. Fuji.

Hinanokinchaku (princess purse) is a small plant only 2-10 cm high whose fruit looks like a purse.

Arinotōgusa (*Haloragis micrantha*) (EN(Y))

This relatively unknown plant is a member of the *Haloragaceae* family and is distributed throughout Japan. Its presence has been confirmed in three locations in Yamanashi Prefecture, and one of these is the northern foot of Mt. Fuji. Because the number of individuals is so small, the plant ranks high on the Yamanashi Red List. The tiny reddish-orange flower blooms from July to September.

Arinotōgusa flower

Grassland

Funabarasō
(*Vincetoxicum atratum*) (VU(J)/EN(Y)/ N-Ⅲ(S))

A member of the *Asclepiadoideae* family ※. This plant is distributed widely in the grasslands of Japan, but its numbers are few, and in Yamanashi, its habitat is restricted. The flower blooms in June and July.

Kaijindō (*Ajuga ciliata*) (VU(J/Y)/EN(S))

This plant is indigenous to Japan, and is found in Kyūshū as well as to the north of Honshū's Chūbu region. The *kai* in **kaijindo** is what the region now encompassed by Yamanashi Prefecture was once called. Fortunately we can see this plant quite often on and around the northern foot of Mt. Fuji.

Funabarasō bud and black-violet flower

※ According to the APG (Angiosperm Phylogeny Group) plant classification system

Kaijindō bud and flower

Murasaki (*Lithospermum erythrorhizon*) (EN(J/Y/S))

Because its root was once used to make a violet dye, this plant is called **murasaki** (violet) despite the fact that its flower is white. This plant can be seen in only one location at the northern foot of Mt. Fuji.

Hanahatazao (*Dontostemon dentatus*) (CR(J/Y/S))

Found in Japan in the Chūbu region and southern part of the Tōhoku region, this winter annual lives in sunlit soil mixed with gravel and blooms from June to August. The color of its petals is reddish violet.

Murasaki flower

Murasaki root

Hanahatazao flower

I. Current status of plants and animals living in grassland environments

D. Endangered grassland insects

① Grassland *lycaenid* butterflies that have a symbiotic relationship with ants

Ants stop parasitic fly from approaching larva.

Facultative symbiosis
The larvae of this group can develop without ants, but ants provide protection from predators in exchange for honeydew supplied by the larvae.

Miyama-shijimi (Reverdin's blue)
(*Plebejus arygyrognomom*) (EN(J)/Y)/VU(S))

Symbiotic ants: *Camponotus japonicus, Formica yessensis, Formica japonica, Pristomyrmex punctatus*

This species can be seen at only limited locations at the northern foot of Mt. Fuji.

Asama-shijimi
(*Plebejus subsolana*) (EN(J)/VU(Y)/CR(S))

Symbiotic ants: *Lasius japonicus, Camponotus japonicus, Camponotus obscuripes*

This species lives in multiple locations at the northern foot of Mt. Fuji and surrounding areas.

Hime-shijimi (Silver-studded blue)
(*Plebejus argus*) (NT(J)/VU(Y)/EN(S))

Symbiotic ants: *Lasius japonicus, Formica japonica*

This species used to be seen widely at the northern foot of Mt. Fuji, but its habit and numbers are in decline.

Obligate symbiosis
The larvae of this group cannot grow unless they are fed by ants or feed on ant larvae.

"Foster" type symbiosis
The larvae initially feed on the dew secreted from aphids, but in their middle stage they are carried by the symbiotic ants to the ants' nest, where the ants feed them mouth to mouth in exchange for the honey dew secreted from their backs.

Kuro-shijimi
(*Niphanda fusca*) (EN(J/Y)/VU(S))
Symbiotic ants: *Camponotus japonicus*

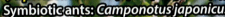

Resting male

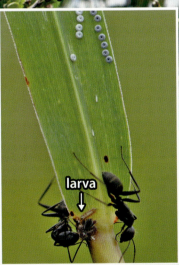
Resting female

Eggs and larva

This species can be seen at only limited locations at the foot of Mt. Fuji.

Predatory symbiosis
The larvae of this species initially feed in the flower bud of *Sanguisorba officinalis*, but in their middle stage they are carried by the symbiotic ants to the ant's nest, where they eat the ants' larvae.

Goma-shijimi
(*Maculinea teleius*) (CR(J/Y/S))
Symbiotic ants: *Myrmica kotokui*

Courtship flight

Female laying egg in the bud of *Sanguisorba officinalis*

This species can be seen at only limited locations at the foot of Mt. Fuji.

② Grassland skipper butterflies

Upper side

Upper side of male

Upper side of female

Underside of female

Silver straight line on the underside of hind wing

The upper side is blackish brown.

Chamadara-seseri (Maculatus skipper)
(*Pyrgus maculatus*) (EN(J)/CR(Y)/EX(S))

The larvae eat *Potentilla freyniana* or *Potentilla fragarioides*, and the adults favor grasslands with patchy bare spots. This kind of environment has recently been shrinking, so the species barely survives at limited locations in Japan, including the foot of Mt. Fuji.

Aka-seseri (Oriental chequered darter)
(*Hesperia florinda*) (EN(J/Y)/EX(S))

The larvae eat *Carex lanceolata* and the adults favor large grasslands with short plants. A dead specimen was once found at an altitude of 2,900 meters in the Fujinomiya Ōsawa Valley, indicating that this species is highly mobile. This species can be seen at only limited locations at the foot of Mt. Fuji.

Gin'ichimonji-seseri (Silver-lined skipper)
(*Leptalina unicolor*) (NT(J)/VU(Y)/N-II(S))

Because its larvae eat *Miscanthus sinensis*, this species used to be distributed widely in *satoyama* environments throughout Japan, but now it is in decline. Fortunately, however, this species continues to thrive at the northern foot of Mt. Fuji because of its many *Miscanthus sinensis* colonies and expansive fallow.

Hoshichabane-seseri (Japanese scrub hopper)
(*Aeromachus inachus*) (EN(J/Y)/CR(S))

The larvae eat *Spodiopogon sibiricus*, and the adults favor grasslands and scattered woods. These environments are in decline all over Japan, and the species can be seen at only limited locations at the foot of Mt. Fuji.

Brown-colored base with star-like black and white dots on underside

Brown-colored base with white dotted line on upper side

Male

Female

Egg

Sujigurochabane-seseri
(*Thymelicus leoninus*) (NT(J)/EN(Y)/EX(S))

The larvae eat *Elymus tsukushiensis*, and the adults favor riverside environments. At the foot of Mt. Fuji these environments are rare, so this species continues to survive in only certain locations. The closely related species **herigurochabane-seseri** *Thymelicus sylvaticus* is also in decline.

③ White and *nymphalid* butterflies that live in grasslands and scattered woods

Kimadara-modoki (Pseudo-labyrinth)
(*Kirinia fentoni*) (NT(J)/VU(Y) /N-Ⅱ(S))

This species has yellow spots (*kimadara*) on both sides of its wings. It favors scattered woods and often rests with closed wings on tree trunks. The larvae eat *Miscanthus sinensis* or *Carex lanceolata*, but they do not live everywhere these plants are found, and their habitat at the foot of Mt. Fuji is extremely limited.

Upper side of male

Upper side of female

Underside of female

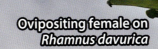
Ovipositing female on *Rhamnus davurica*

Last instar larva on the leaf of *Rhamnus davurica*

Yama-kichō (Brimstone)
(*Gonepteryx rhamni*) (EN(J)/NT(Y)/VU(S))

The larvae eat *Rhamnus davurica*, so the adults favor grasslands and scattered woods. The adults that emerge in August pass the winter and lay eggs in June to July, so the adult stage of this species is quite long. Perhaps owing in part to this life cycle, numbers of this species have recently been in decline at the foot of Mt. Fuji.

Male sucking nectar on *Dianthus superbus*

Grassland

The larvae of both the Northeast-Asian wood white and *Zygaena niphona* (a moth that flies during

Himeshirochō (Northeast-Asian wood white)
(*Leptidea amurensis*) (EN(J)/VU(Y/S))

The larvae of this butterfly eat *Vicia amoena*, but it is difficult to find them in their first and second stages. The adults favor open grassland. The reasons are unclear, but in recent years, the places where they can be found have sharply declined.

Courtship behavior of spring-type female and male (female is above)

Resting summer-type male

Last instar larva resting on *Vicia amoena*

Member of the fritillary group. The wings of butterflies in this group have markings that resemble those of a leopard.

Hyōmonchō (Marbled fritillary)
(*Brenthis daphne*)
(VU(J/Y)/CR(S))

The larvae eat *Sanguisorba officinalis*, and the adults favor open grassland. For reasons that are unclear, the habitat of this species has been in sharp decline at the foot of Mt. Fuji.

Upper side of male

Underside of female

Upper side of female

the day) eat *Vicia amoena* plants.

Adult sucking nectar

Last instar larva resting on *Vicia amoena*

A group of last instar larvae (indicated by arrows)

Benimon-madara (*Zygaena niphona*) (NT(J)/VU(Y))

The larvae produce toxic cyanide compounds in their bodies. The adults contain cyanide as well, giving them a bright red aposematic coloration. This species lives in certain locations in the grasslands at the foot of Mt. Fuji, which is situated at the southern border of their distribution in Japan.

④ Grassland moths that fly during the day

Uraginsuji-hyōmon (Eastern silverstripe) (*Argyronome laodice*) (VU(J)/NT(Y/S))

The larvae eat certain species of violet, and the adults live in *satoyama* grasslands and scattered woods. For some reason it has become extremely difficult in recent years to observe this species. In 2007 it was registered in the Red List of Japan's Ministry of the Environment, and its numbers are declining at the foot of Mt. Fuji.

Sukiba-hōjaku (*Hemaris radians*) (VU(J/Y))

This moth has transparent wings and is shaped like a jet plane. Its larvae eat Japanese honeysuckle and *Patrinia scabiosifolia*. Limited to *satoyama* grasslands, the species is in decline throughout Japan, and in 2012 it was listed by the Ministry of the Environment as an endangered species. It has been found at only certain locations in the grasslands at the northern foot of Mt. Fuji.

Upper side of male

Female

Hovering adult

The *Camponotus japonicus* ants (marked PPG and YYY) that fostered Reverdin's blue "O-1" from its larval stage to adulthood

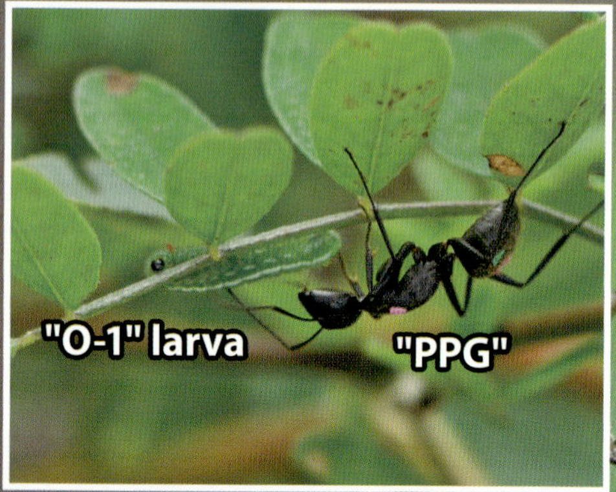

September 1

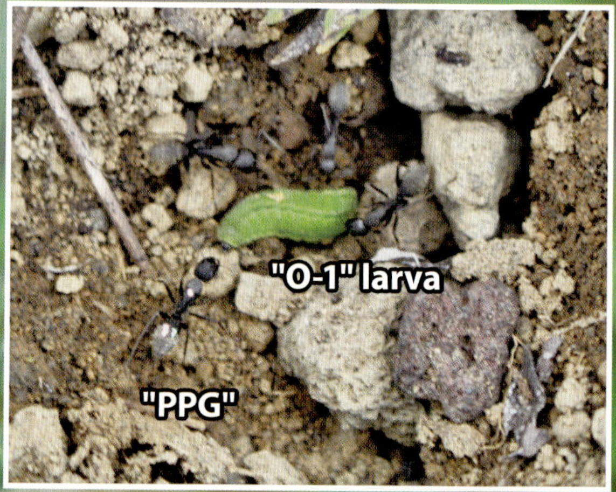

September 10

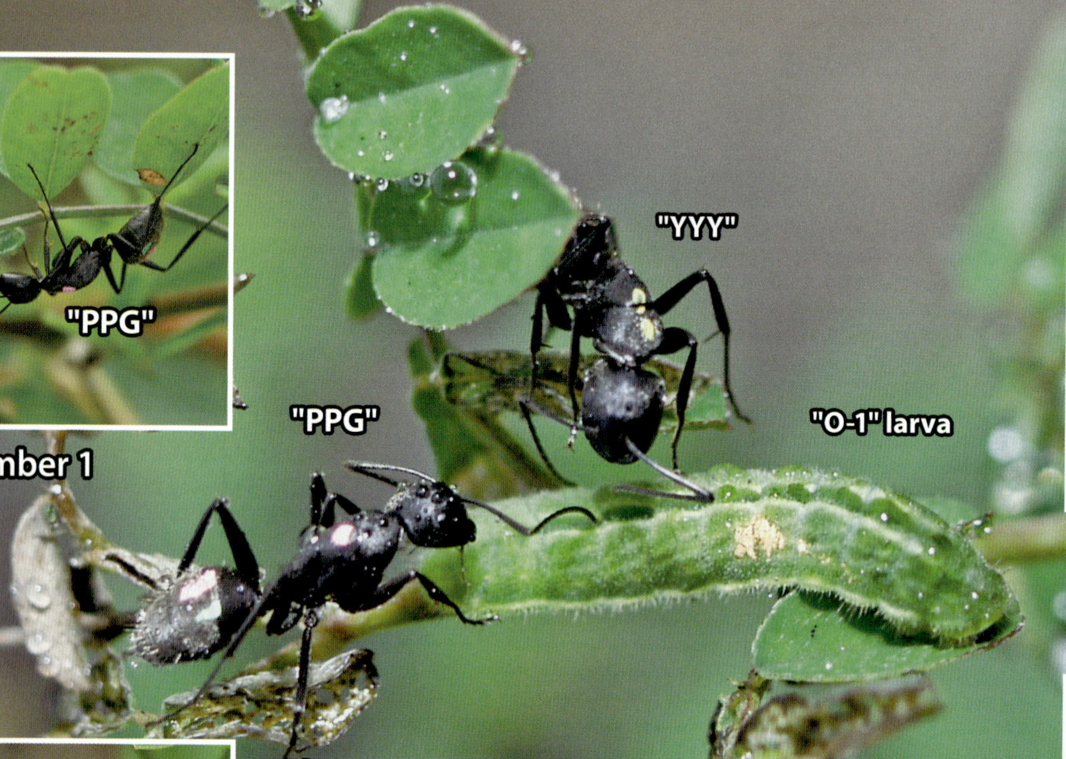

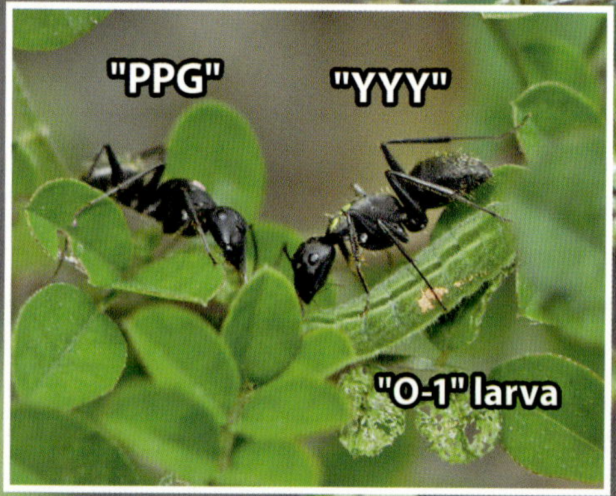

September 6

The *Camponotus japonicus* ants—one marked with two pink dots and one green dot (PPG) and the other with three yellow spots (YYY)—were observed fostering "O-1" (an orange-marked larva of Reverdin's blue) from August to October in 2003. Ants and butterflies are usually enemies, but these species have created a symbiotic relationship, giving us an impressive example of the mysterious workings of nature.

September 7

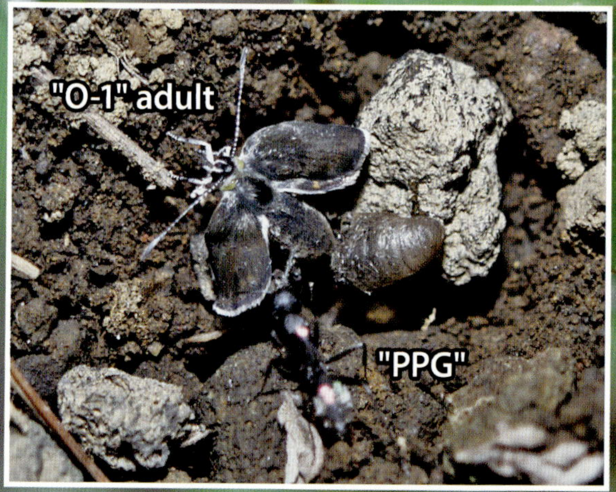

September 30

⑤ Grassland long-horned beetles

Asa-kamikiri (*Thyestilla gebleri*)
(VU(J)/NT(Y)/N-Ⅲ(S))

This species is called **Asa-kamikiri** (*kamikiri* means long-horned beetle) because it was once thought that its larvae only fed on hemp (*asa*) and related species. We now know that they also eat thistle, although this does not mean they are found wherever thistle grows. This species has been observed at only certain locations in the grasslands on and around Mt. Fuji.

Acalolepta degenera resting on the edge of an *Artemisia japonica* leaf

Oblique view of **asa-kamikiri**

Resting on a thistle leaf

Himebirōdo-kamikiri
(*Acalolepta degenera*)
(NT(J/Y)/DD(S))

Because its larvae grow up eating the stalk of *Artemisia japonica*, we usually find this species living by rivers. There are no year-round surface rivers at the foot of Mt. Fuji, however, so we can observe this species in only limited locations in the grasslands.

Resting on the edge of Japanese pampas grass

An example of *Acalolepta degenera* sitting on top of a withered leaf. Its camouflage makes it difficult to spot.

I. Current status of plants and animals living in grassland environments

E. Major threatened grassland birds

Ōjishigi (Latham's snipe)
(*Gallinago hardwickii*) (NT(J/Y)/DD(S))

This is a migratory snipe that arrives from Australia in the spring and lives in grassland environments, making it unusual among birds in the sandpiper family. This species inhabits grasslands or wetlands in the northern part of Honshū's Chūbu region, with Mt. Fuji being the southernmost breeding area. There are extensive grasslands at the foot of Mt. Fuji, but not many members of this species live there. Other breeding grounds are quite distant, so we fear this species will disappear if we do not conserve the grassland habitat.

Circling in the sky and singing "zubiyaku-zubiyaku" to claim territory

Ostentatiously spreading tail feathers while diving. Wind hitting the wings makes the sound "ga-ga-ga."

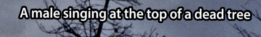
A male singing at the top of a dead tree

A male keeping alert on a telephone pole

An individual coming out from a bush

Akamozu (Brown shrike)
(*Lanius cristatus*) (EN(J/Y)/CR(S))

This species overwinters in India or Southeast Asia and migrates to Japanese woodlands in May. Recently its breeding sites as well as the number of individuals have been in critical decline, making the foot of Mt. Fuji an important breeding ground. Like other shrikes, this species needs a composite habitat consisting of short trees and grasslands. Commercial development and the abandonment of cultivated lands seem to be the main reasons for the decline of both the habitat and numbers of this species.

Adult male

Young bird

Family of nine birds gathered together

Adult male guarding its territory

Adult feeding fledglings

Yotaka (Grey nightjar)
(*Caprimulgus indicus*) (NT(J)/VU(Y/S))

This summer visitor overwinters in Southeast Asia and migrates to woods or grasslands from the foot to the mountain zone of Mt. Fuji to breed. In recent years its population has been declining sharply. These nocturnal birds feed mainly on flying insects such as moths during the night and rest on tree branches during the day. Because they are hard to spot, their numbers are uncertain. However, in the past, whether in the lowlands or mountain zone, one could often hear them cry "kyo-kyo-kyo-kyo," whereas now one usually cannot. The species therefore seems to be in sharp decline.

Mother bird sitting still while away from the nest

Flying adult female

Mother bird feigning injury three weeks later (young birds likely nearby)

Eggs laid on the ground (enlarged photo)

Eggs laid on the ground

Fledglings ten days after hatching (see eggshell to the right)

Grassland

Hayabusa (Peregrine falcon)
(*Falco peregrinus*) (VU(J/Y/S))

The **hayabusa** (peregrine falcon) is about the same size as a crow, with long wings that taper sharply to a point. The underside is white while the head is black, and it is distinguished by whisker-like black patches. We can see this bird mainly along the seashore during the breeding season and in inland cultivated fields or grasslands in winter. They can be found only in a limited area at the base of Mt. Fuji in spring. Because their numbers are in decline and they feed mainly on mid-sized birds, it is important to monitor their feeding conditions.

Misago (Osprey) (*Pandion haliaetus*) (NT(J)/DD(Y) /N-Ⅲ(S))

This bird is about the size of a kite, and the underside, breast, and front of its wings are pure white. It mainly feeds on fish by diving from the air and catching them at the surface of the water. Small in number, it lives primarily along the seashore, but it can sometimes be observed at lakes, ponds, and rivers. This species has been found at Lake Yamanakako and Lake Kawaguchiko.

Flying misago

The peregrine falcon is distinguished by the shape of its wings.

Circling falcon about to dive (April 25, 2010)

Ko-chōgenbō (Merlin)
(*Falco columbarius*) (NT(Y)/N-Ⅲ(S))

Similar in size to a dove, this species is a rare winter visitor. It mainly feeds on small-sized birds in grasslands or cultivated fields, catching them in rapid flights. We can sometimes observe this species along with flocks of small birds at the northern foot of Mt. Fuji.

Resting on a withered tree

Grassland

In flight

Keashi-nosuri
(Rough-legged buzzard)
(*Buteo lagopus*) (Un)

This species is about the same size as a kite. It closely resembles the **nosuri**, but is whitish and has feathers on the base of its legs. This rare winter visitor can be observed at the foot of Mt. Fuji during its migration. Flying low over wide-open spaces such as grasslands or cultivated fields, it hunts mice and other prey.

①

②

③

Hovering, diving, and stopping in mid-air with spread wings (①～③)

Haiiro-chūhi
(Hen harrier)
(*Circus cyaneus*) (DD(Y)/N-Ⅱ(S))

This bird is just a little smaller than a crow, and adult males are entirely gray in color. This uncommon winter visitor hunts prey in grasslands by flying slowly from several to ten meters above the ground. In the winter it hunts mice and small birds.

Circling flight of adult female

Flying adult male

Ko-mimizuku
(Short-eared owl)
(*Asio flammeus*) (NT(Y)/EN(S))

The ear tuft (a feather that looks like an ear) of this species is short and hard to see. This rare winter visitor inhabits grasslands, and is difficult to see because it usually rests in bushes during the day and feeds on mice and other prey after dusk.

Resting on a tree

Male resting on a rock

Ō-mashiko (Rosefinch)
(*Carpodacus roseus*) (DD(Y))

This species is a little bigger than a sparrow, and adult males are entirely pinkish-red in color. The number of these wintering birds is small, and their population fluctuates from year to year. They pick at the fruits of shrubs and eat the seeds of *Polygonaceae* and *Poaceae* plants in and around woodlands from the foot to mountain zone of Mt. Fuji.

Female

Foraging ō-mashiko

Asian rosy finch flying with Fuji in the background

Hagi-mashiko
(Asian rosy finch)
(*Leucosticte arctoa*) (Un)

This bird is similar in size to a sparrow, and males are entirely reddish-violet in color. This winter visitor lives in flocks and usually feeds on the seeds of bush clover, *carex*, and smartweed plants on cliffs and in grasslands.

A flock looking for food on the ground

Male Asian rosy finch

I. Current status of plants and animals living in grassland environments
F. Threatened grassland mammals

Kaya-nezumi (Harvest mouse)
(*Micromys minutus*) (N(Y)/NT(S))

With a body length of 50-80mm and weighing around 10 grams, this is nearly the smallest mouse in the world. This species lives in the *Miscanthus* grasslands of the *satoyama* in and around Mt. Fuji.

Mouse coming out of a *Miscanthus sinensis* bush

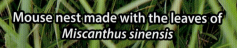
Mouse nest made with the leaves of *Miscanthus sinensis*

Mouse nest viewed from its entrance

This species is known to make a nest like a ball by gathering the leaves of *Poaceae* plants. Ground nests and half-underground nests have been found in the Nashigahara grasslands at the northern foot of Mt. Fuji (Watanabe and Kitagaki, unpubl.). Human activities such as burning are needed to maintain grassland environments.

The grasslands in which the harvest mouse lives are being lost. Registered as a "declining species" in Kanagawa Prefecture's Red Data Book, the mouse is feared to be in sharp decline.

Ground nest found after burning

Nest found in a grassland near the river

Half-underground nest found after burning

Nest found in a biotope at the foot of the mountain

Half-underground nest found two weeks after burning

II. Current status of plants and animals living in *satoyama* woodland environments

A. Why are the *satoyama* woodland environments important?

A temperate monsoon climate with four seasons and plentiful rain plays an important role in the maintenance of coppices and other *satoyama* woodland environments. The broad-leaved deciduous trees characteristic of this climate have a tendency to sprout new growth from their roots after being cut. This growth eventually turns into new trees, a phenomenon we call *bōga kōshin* (see next page).

The photos on the right show the following: 1) an oak tree over 100 years old that was 25 meters high and 80 centimeters in diameter, 2) new growth in the spring the year after the tree was cut down, and 3) growth reaching a height of about one meter two years after cutting. Supporting this kind of growth is a climate with abundant rain and four seasons. In the past, people would take advantage of this by planting broad-leaved trees like **konara** (an East Asian species of oak) to produce firewood and charcoal. The figure on the next page shows the maintenance of coppice through periodic cutting.

As mentioned earlier, the cutting of plantation forests produces valuable grassland that then progresses into scattered woods and eventually new forest. Transitional plantation forests provide important *satoyama* woodland habitat for plants and animals.

In recent years, however, many coppices have been abandoned because of the decreasing need for firewood and charcoal, and plantation forests have gone uncut as well. The result is that many plants and animals living in coppice or plantation forests face extinction due to habitat loss. The current status of these plants and animals is covered in the pages that follow. Together we should consider how to conserve these endangered plants and animals.

1) June 1, 2012

2) May 22, 2013

3) July 12, 2014

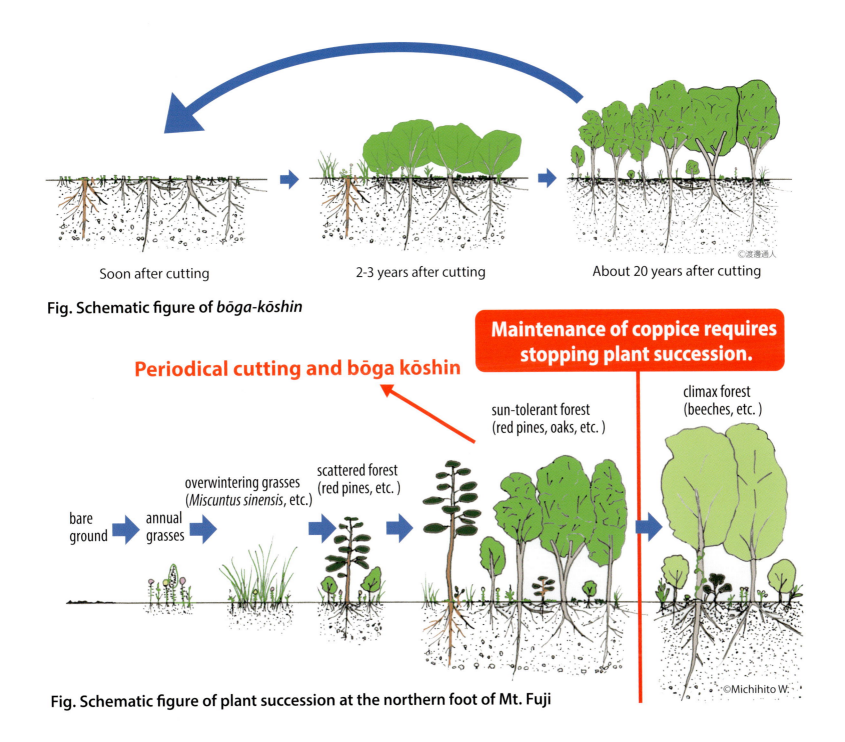

Fig. Schematic figure of bōga-kōshin

Fig. Schematic figure of plant succession at the northern foot of Mt. Fuji

II. Current status of plants and animals living in *satoyama* woodland environments

B. Typical threatened woodland plants

Because of the changes to *satoyama* environments, an increasing number of plants have become threatened not only in grasslands but also woodlands throughout Japan, including at the northern foot of Mt. Fuji.

Suzumushisō (*Liparis makinoana*) (EN(Y)/VU(S))
Fugaku-suzumushisō (*Liparis fujisanensis*) (VU(J)/CR(Y)/EN(S))

The bloom of this species is shaped like the wing of a **suzumushi** (bell cricket). In the past, it was common to find the plant in *satoyama* forests, but it is now in sharp decline in Yamanashi Prefecture.

Jigabachisō (*Liparis krameri*) (VU(Y))

The flower of this *Orchis* species looks like a **jigabachi** (a kind of bee). *Orchis* plants such as these are declining sharply in Yamanashi Prefecture because people like to pick them. This species is now rarely seen at the northern foot of Mt. Fuji.

On the floor of stable broadleaf forests we sometimes find saprophytes.
Saprophytes survive by consuming bacteria in the ground.
Stable broadleaf forests have been decreasing, so some saprophytes have been designated threatened plants.

Sakane-ran (*Neottia nidus-avis*) (VU(J)/ EN(Y/S))

This plant is said to be named **sakane-ran** because it is an *Orchis* (*ran*) whose roots (*ne*) appear to grow upside down (*saka*). It is found in one region at the northern foot of Mt. Fuji.

Oninoyagara (*Gastrodia elata*) (VU(Y))

This plant is called **oninoyagara** because it is said to look like an arrow (*yagara*) used by a demon (*oni*). It is found in some areas at the northern foot of Mt. Fuji, but there is little information about its distribution in Yamanashi Prefecture.

The rhizome is called **tenma**, with the blue variety called **ao-tenma**.

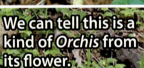

The root of **sakane-ran**

We can tell this is a kind of *Orchis* from its flower.

Flower seen from above

Yama-shakuyaku (*Paeonia japonica*) (NT(J/Y/S))
The flower of this plant is white, and its name means "peony which blooms in the mountains." It is declining throughout Japan, but fortunately, we can still see it quite often at the northern foot of Mt. Fuji.

Benibana-yama-shakuyaku (*Paeonia obovata*) (VU(J)/CR(Y/S))

This species is distributed widely over the Japanese islands, but because its population is small, it is rarely seen. At the northern foot of Mt. Fuji, its numbers have shrunk dramatically. The color of its flower is crimson and violet.

Ko-atsumorisō (*Cypripedium debile*) (NT(J)/EN(Y)/VU(S))

This perennial plant of the *Orchidaeceae* grows on the floor of mature forests. In all parts of Japan, its habitat and population are scarce, and it is rarely seen at the northern foot of Mt. Fuji.

The flower of the **benibana-yama-shakuyaku** is crimson and violet whereas that of the **yama-shakuyaku** is white, but both species produce similar looking fruit (see upper left) that splits open on its own. The unripe seeds turn red and the ripe ones turn dark blue.

The flower is the same size as a fingertip.

II. Current status of plants and animals living in *satoyama* woodland environments

C. Representative threatened insects

Male adult with wings shining in the sunlight

Ōmurasaki (Great purple emperor)
(*Sasakia charonda*) (NT(J)/N(Y)/N-III(S))

Because there are few wild hackberry trees (**enoki**) at Fuji's northern foot, the larvae of this insect primarily eat *Celtis jessoensis* (ezo enoki). The larvae turn dark brown during the winter, which they spend among dead leaves on the ground. In the spring, they turn green after climbing up into the fresh green leaves, where they develop into pupae and then emerge as adults in July or August. The adults like sap, so their main habitat is coppice.

Observing the insect's growth in the field

Two, fourth-instar larvae wintering among dead leaves

Last-instar larva

Pupa (left) and pupal shell (right) under a leaf

A male adult that has just emerged

Adult eating pollen of **yama-shakuyaku**

Futasuji-katabiro-hanakamikiri
(*Brachyta bifasciata*) (NT(Y))

The name of this species can be translated into English as "a long-horned beetle with two stripes and wide shoulders that gathers flowers." Because of its long name, it has been nicknamed simply **kimaru** (meaning "yellow and round") by insect lovers. This species likes to eat the pollen and petals of **yama-shakuyaku** and **benibana-yama-shakuyaku**. It oviposits in the stalk at ground level, and it is said that its larvae eat root starch. As shown earlier, yama-shakuyaku and benibana-yama-shakuyaku are now threatened, so it is rare to see this long-horned beetle as well. Fortunately, at the northern foot of Mt. Fuji, we can often find **yama-shakuyaku**, but **benibana-yama-shakuyaku** is very rare, so only if you are lucky will you come across this flowering species.

II. Current status of plants and animals living in *satoyama* woodland environments

D. Representative threatened birds

Ō-taka (Northern goshawk)
(*Accipiter gentilis*) (NT(J/Y/S))

Similar in size to a crow, this hawk breeds in the forest and preys on mid-sized birds such as pigeons and small mammals such as rabbits. Although it is considered a "nearly threatened" species by the Ministry of the Environment and Yamanashi and Shizuoka prefectures, in recent years, this forest-dwelling bird has been spotted increasingly in urban areas. Neither this population increase nor the bird's hunting behavior in the forest is well understood. At the northern foot of Mt. Fuji, the number of this species is limited, but it can still be widely seen.

Young ō-taka resting on a branch

Adult ō-taka flying over the forest

Young ō-taka perching on a branch

Hai-taka (Sparrowhawk)
(*Accipiter nisus*) (NT(J)/VU(Y/S))

This hawk is similar in size to a domesticated dove. It breeds in the mixed forests of the montane zone, and it can also be seen in lowland forests in autumn and winter. It preys mainly on small birds, and although there is not much data about the number of inhabitants, it seems to be very small. It is sometimes seen along with groups of small birds at the northern foot on Mt. Fuji.

Adult hai-taka resting on a branch

Adult hai-taka chasing a flock of hagimashiko (Asian rosy finch)

Fukurō (Ural owl)
(*Strix uralensis*) (NT(Y/S))

This owl lives in forests in both the lowlands and mountainous areas. In the daytime it often rests in a nesting cavity, and at night it flies without making a sound and attacks small mammals such as mice, small birds, and insects. It nests mainly in the cavities of big trees, so the decline in the number of big trees in recent years has endangered their breeding. Fortunately, in the breeding season, we can still hear it sing "**gorosuke-hō-kō**" in a considerably wide range at the northern foot of Mt. Fuji.

Adult catching a Japanese grass vole

Resting on an electric wire in the morning

Young birds in a nesting cavity

Stopping on a cage and looking across the pond

Woodland

Sankōchō (Japanese paradise flycatcher)
(*Terpsiphone atrocaudata*) (NT(Y/S))

This bird is about the same size as a sparrow, but the male's tail in the breeding season is particularly long, reaching about 30 centimeters. This flycatcher is distinguished by its song, **"tsuki-hi-hoshi hoi-hoi-hoi."** It migrates as a summer visitor and lives in dark forests, where it hunts flying insects. The population is in decline.

Parents feeding their young

Toratsugumi (White's thrush)
(*Zoothera dauma*) (NT(Y))

This thrush is a little bigger than a grey starling. Its body is yellow-brown with black, scaly spots that make it look like a tiger. It lives in forests in both low-mountain and subalpine areas, and at night feebly sings **"hi-i hi-i,"** which was once thought to be the cry of a mythical chimera known as a *nue*. It often pushes through dead leaves looking to eat earthworms and insects living in the ground. The population has been in decline at the northern foot of Mt. Fuji in recent years.

Individual resting on a branch

Sanshōkui (Ashy minivet)
(*Pericrocotus divaricatus*) (VU(J)/NT(Y)/EN(S))

The population of this summer visitor, which is similar in size to a wagtail, is in decline. Its characteristic song sounds like "hi-ri-ri hi-ri-ri," and it lives in broad-leaved forests in both low-mountain and subalpine zones. It forms small groups and preys on scarabs and other insects.

Adult resting on a branch

Nojiko (Japanese yellow bunting)
(*Emberiza sulphurata*) (NT(J/Y))

This bunting migrates in spring from southeastern China and the Philippines, and nests in the sunlit forests of both the *satoyama* and mountain zones, where it feeds on insects and seeds. It appears to breed only on the Japanese island of Honshū. Although rare, this bird can fortunately be seen in a wide range at the northern foot of Mt.Fuji.

Adult at a puddle

Adult carrying an insect

Alert adult stopping on a branch

Buppōsō (Broad-billed roller)
(*Eurystomus orientalis*) (EN(J/Y)/CR(S))

This summer visitor is a little bigger than a grey starling and preys on flying insects in lowland and mountain forests with large trees. The population is decreasing because of the decline in big trees where they nest. It can be found in only one area at the northern foot of Mt. Fuji.

Woodland

Resting male

Female looking back

Ōkonohazuku (Sunda scops owl)
(*Otus lempiji*) (VU(Y)/DD(S))

Slightly smaller than a pigeon, this bird is brown with complex patterns of black and grey, and the feathers on its head look like horns. It is difficult to find because it does not sing with a loud voice and is nocturnal, but the population is thought to be limited. Preying on small birds, mice, and other animals, this species lives in forests with big trees surrounding shrines and temples. These forests have been decreasing in recent years, so the population of this bird seems to be decreasing as well.

Isuka (Common crossbill) (*Loxia curvirostra*) (DD(Y))

This bird is similar in size to a meadow bunting, and the male is red. Although it is a winter visitor, there are reports that it has been breeding at the northern foot of Mt. Fuji in recent years. In winter it lives in mountain forests, forming groups ranging from several to ten birds and skillfully using its unique beak to extract pinecone seeds.

Individual at a puddle

Kuroji (Grey bunting)
(*Emberiza variabilis*) (DD(Y))

This bird is about the size of a meadow bunting. In the breeding season it lives in subalpine forests, while in other seasons it inhabits dark forests higher up in the mountains. The population of this species is small, so it is not easy to find. Its song sounds like "hōi chiyo-chiyo," but outside the breeding season, it simply twitters "chi" in a metallic voice. It feeds on insects and spiders in the trees during the breeding season and on seeds in other seasons.

Adult resting on a branch

Renjaku (waxwings) gather in mistletoe to eat their fruit. Mistletoe fruit contains a viscous fluid, so by eating it, these birds increase the stickiness of their excrement, which in turn enables the fruit to attach to tree branches and germinate. As a consequence, the distribution of mistletoe is closely associated with the migratory movements of renjaku.

Ki-renjaku (Bohemian waxwing)
(*Bombycilla garrulus*) (NT(Y))

This is a winter visitor, similar in size to a shrike. It has a crown feather and yellow tips on its tail (the *ki* in **ki-renjaku** means yellow). Its migrating population varies widely, and there are years when it is not found at all. During the winter it moves in groups of several dozen birds. It eats the seeds of mistletoe and other plants.

Resting on a branch

Hi-renjaku (Japanese waxwing)
(*Bombycilla japonica*) (NT(Y))

This winter visitor is slightly smaller in size than the **ki-renjaku** and its tail tip is red (the *hi* in **hi-renjaku** means red). The number of migrants varies widely, so in some years none have been spotted. It often moves in groups of several dozen birds, although sometimes there are more than 100 individuals in a group. It eats the seeds of mistletoe and other plants.

Hi-renjaku perching on a power wire

II. Current status of plants and animals living in *satoyama* woodland environments
E. Threatened mammals

Yamane about to eat mountain grapes

Yamane (Japanese dormouse)
(*Glirulus japonicus*)
(National designated natural monument)
(NT(Y)/DD(S))

The indigenous **yamane** (Japanese dormouse) is distributed throughout the Japanese islands. A nocturnal species, it lives in forests from the low-mountain to subalpine zone and inhabits the *satoyama* at the northern foot of Mt. Fuji. It eats mainly fruit and insects, and hibernates during the winter.

The **yamane** lives in plantation forests and surrounding areas, but it is difficult to survive throughout the year in the forest where the ground is dry, so it prefers mature forests with lots of hiding places and plenty of insects.

Individual resting on the trunk of a Japanese red pine

Individual climbing the trunk of a Japanese red pine

Captured individual from the back

Captured individual from the front

Running after release

II. Current status of plants and animals living in *satoyama* woodland environments

F. Rare amphibians

Non-speckled type

Speckled type

Moriao-gaeru (Forest green tree frog)
(*Rhacophorus arboreus*) (NT(S))

Moriao-gaeru means "bluish-green forest frog." Individuals are scattered in different locations, so it is hard to spot. Around June the frogs gather where there is water and oviposit. Their frothy white egg sacs make them traceable. Waterfront is relatively limited at the northern foot of Mt. Fuji, but this species finds small puddles or ponds where it can oviposit. It is widely distributed, but is rarely seen above 1000 meters. In Yamanashi Prefecture, Nanbu and other municipalities have designated this species a natural monument that needs protection.

Males gathered around a female to make an egg sac

Egg sac near completion

II. Current status of plants and animals living in *satoyama* woodland environments

G. Why are *satoyama* lava flows important?

Mt. Fuji is a living volcano. The last big eruption, the Hōei eruption, took place about 300 years ago, and before then, numerous lava flows originated from various craters. The figure on the prior page depicts in different colors the lava flows that formed in the last stage of New Fuji (2,000 years ago onward). As you can see, many eruptions took place during this period. Flows originating before 2,000 years ago are uncolored. The photo to the right shows that Mt. Fuji consists of many layers of lava and pyroclastic material.

The contents and temperature of each lava flow determine its viscosity and speed, and under the right conditions, some flows create lava caves and lava tree molds (see pages 55 and 56). For large caves such as the Fuji Fūketsu Cave or Saiko Bat Cave, there is a big difference in air temperature and humidity between the entrance and the interior. There is even a significant difference in the case of 10-meter-long lava tree molds, where the temperature inside varies little throughout the year, remaining cool in summer and warm in winter.

Starting on page 58, you will learn about the current status of the various animals that live in lava fields, which we should all work together to conserve.

View of the left side of the Fujinomiya Ōsawa Valley

View of the upper part of the first Hōei crater

Column 4

Kenmarubi lava flows

The **Kenmarubi** lava flows originated in the years following 937 AD from a fissure toward the summit of the northern side of Mt. Fuji. The first lava flow engulfed and burned out dense stands of large, ancient trees that grew in the valley below. As you can see from the figure at the bottom of this page, some trees burned while still standing and others after falling down. This process created what are called "lava tree molds." The **Funatsu Tainai** and **Yoshida Tainai** lava tree molds, both of which have been designated World Cultural Heritage Sites, are complex lava tree molds that were formed by the burning of multiple trees that fell on top of one another.

Few lava tree molds were produced by the second **Kenmarubi** lava flow because, by that point, almost all of the trees in the valley had already burned.

Five types of lava tree molds (schematic figure)

① Well mold

② Stone pillar mold

③ Horizontal mold

④ Leaning mold

⑤ Driftwood mold

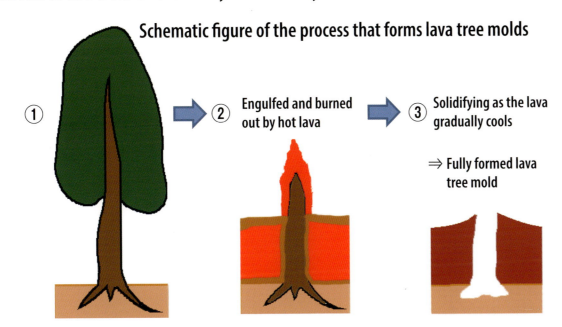

Schematic figure of the process that forms lava tree molds

① → ② Engulfed and burned out by hot lava → ③ Solidifying as the lava gradually cools ⇒ Fully formed lava tree mold

©Michihito W.

Column 5

Aokigahara lava flow

This lava flow originated in 864 AD from a fissure near Mt. Nagao, a parasitic cone located on Fuji's northwestern flank. In some places it is up to 135 meters thick. Well-known lava tunnels (small ones are called lava tubes) include the Saiko Bat Cave, Fuji Fūketsu Cave, Motosu Fūketsu Cave No. 1, Motosu Fūketsu Cave No. 2, Fugaku Fūketsu Cave, and Narusawa Hyōketsu Cave.

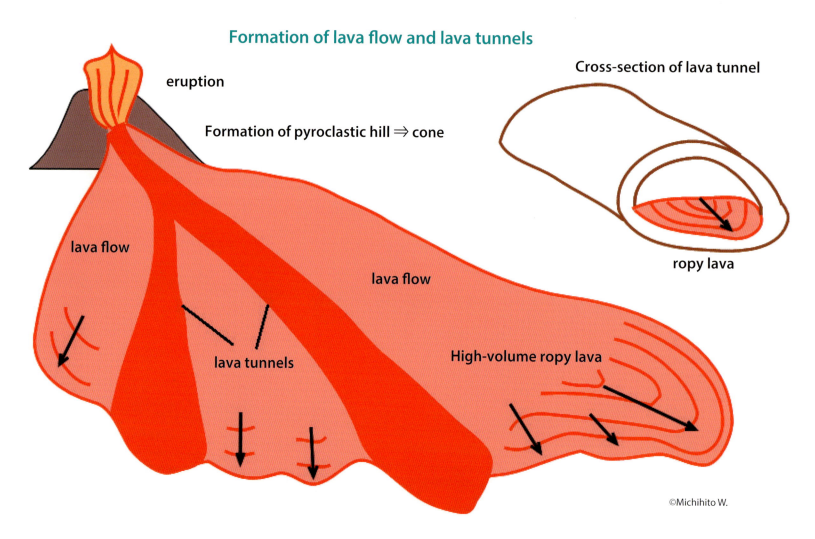

©Michihito W.

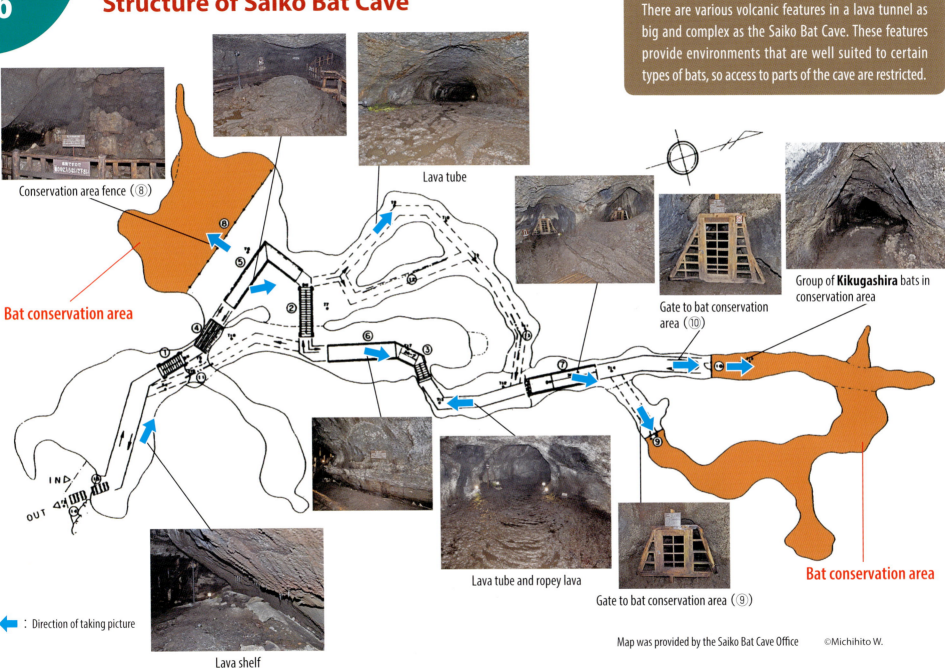

II. Current status of plants and animals living in *satoyama* woodland environments

H. Animals uniquely adapted to living in lava fields

① Bats living in lava caves

Usagi-kōmori (Brown long-eared bat)
(*Plecotus auritus*) (NT(Y)/DD(S))

These bats are distinguished by very long auricles. However, they fold their auricles while resting, so it can be hard to tell them apart from other bat species. In the daytime they hide in the cavities of large trees, but they also rest in caves and houses. They give birth in the early summer.

Kikugashira-kōmori (Greater horseshoe bat)
(*Rhinolophus ferrumequinum*) (N(Y)/NT(S))

These bats are widely distributed in Japan. During the daytime they rest in caves in groups that can reach as many as several hundred individuals. In the caves at the northern foot of Mt. Fuji, they often formed mixed groups with another bat species, Momojiro-kōmori.

Resting adult

Sleeping adult

Adults resting in a group

Two adults resting on the ceiling in the conservation area

Sleeping adult

Tengu-kōmori (Tube-nosed bat)
(*Murina hilgendorfi*) (NT(Y)/DD(S))

This species is indigenous to Japan and has been spotted widely from Hokkaidō to Kyūshū, but there is not much information about it. Although it tends to hide in tree cavities in the daytime, it can also be found in caves. It is thought to prey mainly on forest insects.

Resting individual

Momojiro-kōmori (Japanese large-footed bat)
(*Myotis macrodactylus*) (NT(Y/S))

During the day this bat species hides mainly in caves, and all year long it is common for it to form groups of over 100 individuals. At the northern foot of Mt. Fuji, we can observe how it hunts insects by lakes and rivers after sunset. Its life span is estimated to be more than six years.

Adults sleeping next to each other

Ko-kikugashira-kōmori (Little Japanese horseshoe bat)
(*Rhinolophus cornutus*) (N(Y)/NT(S))

Indigenous to Japan and mainly living in caves, these bats often rest in groups of more than 100 individuals during the day. They prey mainly on small insects while skimming the surface of the water or the ground. They give birth in early summer.

Adults flying out in a line after sunset

II. Current status of plants and animals living in *satoyama* woodland environments

H. Animals uniquely adapted to lava fields

② Insects living in lava caves

Group of **gaganbo** (crane flies)

Madara-kamadouma found in a cave in March

Ice "bamboo shoot" in cave in the middle of summer

Insect on "bamboo shoot"

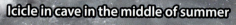
Icicle in cave in the middle of summer

Enlarged view of "bamboo shoot" (reflected in water)

A group of moths

Individual resting on the tree mold ceiling

Usagi-kōmori flying in a shrine

Maemon-ō-namishaku
(*Triphosa sericata*) (Un)

The name of this species means "large *geometrid* moth with spots on its forewings." It has an unusual life cycle in that it emerges in the summer, passes the winter as an adult in a cave, and mates and oviposits the following spring. Therefore, it can be found in lava caves (or tree molds) only at certain times. Because it has a life cycle adapted to the special environment of lava caves, human impacts are of great concern.

Wings fallen on the ground, perhaps due to predation by bats

III. Current status of plants and animals living in *satoyama* waterfront environments

A. Why are *satoyama* waterfront environments important?

Although there is no permanent river running down the surface of Mt. Fuji, there are waterfront environments surrounding the mountain, such as the Fuji Five Lakes, Oshino Hakkai (the eight springs of Oshino), and the Katsura River, which flows from Lake Yamanakako and has several tributaries. As mentioned earlier, Mt. Fuji consists of many layers of lava and pyroclastic materials. Most rainwater sinks through these layers until reaching one that is relatively impermeable, such as one of the Old Fuji mudflows. It then runs on top of this layer until gushing out of the ground as spring water.

The Fuji Five Lakes and Oshino Hakkai are fed by spring water from Fuji and rainwater from nearby mountains. As shown in the photos to the right, the surface level of the lakes fluctuates widely throughout the year, so the wetlands surrounding them are free of plants that tend to grow in cultivated fields or around human habitation.

Unlike algae living in lake water, wetland plants are restricted to a narrow zone at the lake's edge (see the figure on the next page), so the distribution of wetland plants fluctuates with the surface level of the lake. Natural waterfront environments such as lakeshores and riverbanks provide precious habitat for endangered wetland plants.

The surface level of these lakes varies according to season, and wetland plants cannot live along lakeshores or riverbanks hardened by concrete, so these plants can now be found in only limited areas. Many of them are listed as endangered species.

The current status of endangered plants in *satoyama* waterfront environments is covered from p. 64. We should all consider how to protect these plants.

July 3, 2004

July 19, 2007

June 30, 2014

The figure below is a schematic cross section of one of the Fuji Five Lakes.

The southern shores of the Fuji Five Lakes consist of lava, while the northern ones are flanked by mountains made up of much older geological layers dating to the Neogene period (23.03-2.58 million years ago), so the opposite sides of each lake look quite different from one another. This variety of lakeshore environments, one of the distinguishing features of the Fuji Five Lakes, creates an ecological mosaic in which wetland plants can spread their seeds according to changes in lake levels and establish themselves in suitable niches.

Geological variety supports the biodiversity of wetland plants, which in turn sustains the animals that favor waterfront habitats. Thus the Fuji Five Lakes, along with the Katsura River and its tributaries, provide valuable *satoyama* waterfront environments at the northern foot of Mt. Fuji.

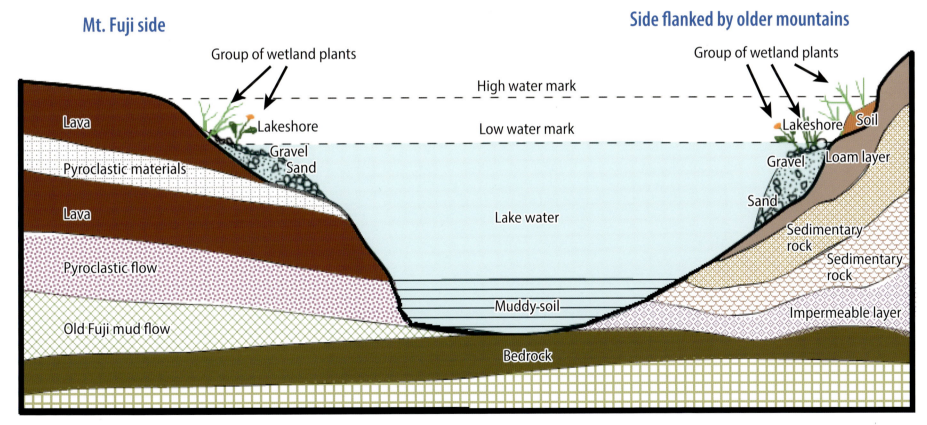

Fig. schematic cross-section of lake

©Michihito W.

III. Current status of plants and animals living in *satoyama* waterfront environments

B. Representative wetland plants that are threatened

Since many of the plants growing in the wetlands are not very noticeable, people's interest in them seems to be low, but there are many plant species that face extinction.

Yamanashi Prefecture has few wetlands, making the waterfront areas at the northern foot of Mt. Fuji especially valuable. The habitat of the wetland plants in these areas is influenced by yearly and seasonal changes in water level.

October 12, 2010

November 21, 2012

October 17, 2013

October 28, 2014

Environmental changes to the habitat of threatened wetland plants

Jōrōsuge (*Carex capricornis*) (VU(J)/CR(Y))
This northern plant has small populations and is sparsely distributed. In Yamanashi Prefecture, it is found only around the Fuji Five Lakes. The population and distribution of this species fluctuate along with water levels.

The spike of **jōrōsuge**
A large colony of **jōrōsuge** was observed on May 31, 2007.

Observations on June 30, 2014 revealed that portions of a **jōrōsuge** colony were recovering, although they could be seen only on occasion because of the changing water level.

Sujinumahari'i (*Eleocharis equisetiformis*)
(VU(J)/EN(Y))

This species is distributed widely in Japan, but its population is small. In Yamanashi Prefecture, it can only be seen around the Fuji Five Lakes.

A member of the **igusa** family with spikes shaped like needles

Takonoashi (*Penthorum chinense*)
(NT(J)/EN(Y)/NT(S))

This species is distributed widely in Japan. In Yamanashi Prefecture it has been found in flooded rice paddies as well as the muddy wetlands around lakes. The Fuji Five Lakes provide important habitats for this species.

It is called "octopus tentacles" because of the shape of its spikes, which have many small flowers.

Waterfront

Kawajisha (Water speedwell)
(*Veronica anagallis-aquatica*) (NT(J/Y))

This species grows on riverbanks and lakesides in the Chūbu region and western Japan. It used to be a relatively common plant, but it has been declining in recent years.

Matsukasa-susuki
(*Scirpus mitsukurianus*) (VU(Y)/N-III(S))

This plant is distributed widely in the islands of Honshū, Shikoku, and Kyūshū. In Yamanashi Prefecture, it is found only around the Fuji Five Lakes and one other wetland area. A closely related species is **ko-matsukasa-susuki**.

Kawajisha flower

The spikes of **matsukasa-susuki** resemble pine cones.

Ko-kitsunenobotan
(*Ranunculus chinensis*) (VU(J)/EN(Y))

This species likes cold climates and its numbers are decreasing nationwide. In Yamanashi Prefecture, it is seen only around the Fuji Five Lakes.

Ko-kitsunenobotan fruit clusters are elongated and elliptical, but those of the frequently seen **kitsunenobotan** are spherical.

As the word *ko* (small) at the beginning of **ko-kitsunenobotan** indicates, the flower of this species is smaller than that of **kitsunenobotan**.

Kayatsuri-suge
(*Carex bohemica*) (EN(J/Y))

The distribution of this species in Japan is unusual in that it lives only in Hokkaidō and in the Fuji Five Lakes region of Yamanashi Prefecture. It grows in the sand and gravel of sunny lakeshores.

A member of the **suge** family whose spike resembles that of *Cyperaceae*

Baikamo
(*Ranunculus nipponicus*) (VU(Y))

This algae species grows in springs and rivers that maintain a stable water temperature throughout the year. In Yamanashi Prefecture it has been found only at the northern foot of Mt. Fuji and in parts of Tsuru City.

Mikuri
(*Sparganium erectum*) (NT(J)/CR(Y)/NT(S))

In Yamanashi Prefecture it grows only along the clear streams found in Oshino Village and around Lake Yamanakako.

Fruit   Flower

It is called **baikamo** (plum blossom algae) because its flowers look like plum blossoms.

III. Current status of plants and animals living in *satoyama* waterfront environments

C. Rare Amphibians · Reptiles

Tonosama-gaeru
(Black-spotted pond frog)
(*Pelophylax nigromaculatus*)
(NT(J)/VU(Y)/NT(S))

It has recently been discovered that Tōkyō-daruma-gaeru and Tonosama-gaeru are two distinct frog species. Yamanashi Prefecture is close to the boundary line between the territories of both species. This frog is in steep decline in the center of the Kōfu Basin, but is sometimes seen in the surrounding mountains. Although not many live at the foot of Mt. Fuji and nearby areas, these frogs can sometimes be seen along rivers at the base of surrounding mountains or in paddy fields with few pesticides.

Individual resting near the river

Shima-hebi
(Japanese striped snake)
(*Elaphe quadrivirgata*) (VU(Y))

As the name suggests, this non-venomous snake has stripes (*shima*). Its population is declining sharply in the center of the Kōfu Basin, and its habitat is now limited to mountain slopes and foothills in the rest of Yamanashi Prefecture, so the prefecture has designated it an endangered species. Fortunately, it is still seen at the northern foot of Mt. Fuji, but its numbers appear to be decreasing. You can readily see this species in the surrounding mountains, but it is rare to find them at the northern foot of Fuji.

Two snakes that may be copulating (1)

Two snakes that may be copulating (2)

Mamushi (*Gloydius blomhoffii*) (Un)

Many people know the name of this poisonous snake, but it is rare to come across one. Until around 30 years ago, this species was frequently spotted in lava fields, but now it is unusual to see them, and one rarely hears of them biting anyone. Although not listed in the Red Data Book of Yamanashi Prefecture, this species is in sharp decline at the northern foot of Mt. Fuji, so we mention it here.

Slithering mamushi

6. Thinking about the future of Mt. Fuji's *satoyama*

I. The problem of invasive plants and animals

① Arechi-uri (Burr cucumber) (*Sicyos angulatus*)
Invasive Alien Species

First discovered in Shizuoka Prefecture in 1952, it is now distributed throughout Japan except Hokkaidō and the southern part of Kagoshima Prefecture. This vine grows extremely quickly, with each plant producing 400 to 500 seeds. Japan's Ministry of the Environment officially designated it an Invasive Alien Species out of concern that its tremendous growth along rivers is displacing indigenous plants (from Ministry of the Environment home page).

In 2010, burr cucumber was discovered in certain locations along Lake Kawaguchiko. Fearing that it would out-compete endangered wetland plants indigenous to the region, we have made efforts over the past eight years to keep the species from spreading. These have succeeded in limiting the plant's growth, so please cooperate by not spreading its seeds.

Sprout in June

Immature fruit

Mature fruit

Mature seed

Two colonies of burr cucumber, each about 100 square meters in size, were discovered in 2010 (October 12, 2010).

In the following year, there was little change to the inland colony, but the one on the lakeshore had disappeared, probably because of the rising level of the lake (Oct. 14, 2011).

Mature colony by the lake (October 14, 2011)

The colony was prevented from spreading to a lower location (October 14, 2011).

No burr cucumber was spotted in the year the area was mowed (October 17, 2013).

There was no new growth the following spring (June 30, 2014).

A resurgence of the colony was prevented (October 28, 2014).

② **Araiguma** (Common raccoon) (*Procyon lotor*)   Invasive Alien Species

First spotted in Aichi Prefecture in 1962, the common raccoon has expanded its distribution nationwide. Because it eats a wide variety of foods, including small mammals, fish, birds, amphibians, insects, vegetables, fruits, and grains, there is concern about its impact on indigenous species of plants and animals. Japan's Ministry of the Environment has therefore officially designated it an Invasive Alien Species (from Ministry of the Environment home page).

At the northern foot of Mt. Fuji, this raccoon has been observed from about the year 2000 around Lake Kawaguchiko and mountainside country homes. Individuals were likely reared and then released by people inspired by the popular television animé "Araiguma rasukaru." The species has been found in a wide range, including both *satoyama* and forested areas, so there is concern that it will continue to expand its distribution and numbers. The photo to the left, taken by a motion sensor camera, shows a raccoon making one of its periodic visits to a water hole. Common raccoons are difficult to trap, so it is hard to control their numbers.

③ **Ōkanada-gan** (Canada goose) (*Branta canadensis*)   Invasive Alien Species

It is estimated that several dozen Canadian geese live around Mt. Fuji. Closely related to the **shijūkara-gan**, which appears as a Critically Endangered Species in the Ministry of the Environment's Red List, this bird has been officially designated an Invasive Alien Species. Because the population of this highly reproductive bird is increasing, there are fears that it will crossbreed with the **shijūkara-gan** and eat agricultural products. Nowadays the population of this species is in decline around Lake Kawaguchiko due to efforts to catch the birds and to replace real eggs with fake ones in order to interfere with their breeding.

④ **Yokozuna-sashigame** (*Agriosphodrus dohrni*)   Invasive Species

This insect was first recorded in Kyūshū in the 1930s and it invaded the Kantō region in the 1990s (from the Invasive Animals and Plants Data Base of the National Institute for Environmental Studies). Its distribution has been expanding into northern areas because of global warming. It had not been recorded in Yamanashi Prefecture until 2008, when it was found in the southern part of the prefecture (larvae shown in photo to the left). Recently this species has been observed widely in Nirasaki City and other parts of the Kōfu Basin.

At the northern foot of Mt. Fuji, it was observed at a location around 1000 meters in altitude (see photo to the right). The good news is that is has not been seen much recently, but there remains concern that its distribution will increase along with global warming.

II. Things to consider when experiencing the nature of Mt. Fuji

A. Mt. Fuji as a national park

Everywhere around the fifth station, including walking trails, is in a Special Protection Zone.

Mt. Fuji is at the center of the Mt. Fuji region of Fuji-Hakone-Izu National Park. New classifications for national park land were instituted in 1996 (see next page), and the regulations of the Natural Parks Law put limits on commercial development.

The collecting of animals, plants or rocks is prohibited in the Special Protection Zones of national parks. However, most people are ill informed about these areas, whose borders are difficult to identify. Before 1996 you could collect cowberries, mushrooms or **oniku** (*Boschniakia rossica*)—now an endangered plant in Yamanashi Prefecture—if you had permission from the Onshirin-kumiai (an organization that protects traditional common land rights for local communities on the northern side of Fuji). Since 1996, however, there have been cases of people entering the Special Protection Zone to collect these items without first having received permission. Almost all of Fuji from the fifth station to the summit is located in a Special Protection Zone, but how many people climbing to the top of Fuji, the most summited mountain in the world, are aware that the climbing trails are located in a Special Protection Zone of a national park?

In order to preserve the nature of Mt. Fuji, a World Cultural Heritage Site, we need to figure out how to raise the awareness of climbers and other visitors about Fuji's place within Fuji-Hakone-Izu National Park.

Regulation of development

One of the aims of national parks according to the Natural Parks Law is to protect beautiful natural landscapes. In order to fulfill this aim, the law regulates such activities as constructing buildings, cutting trees or bamboo, and collecting rocks, animals, and plants. The National Park Plan (a set of conservation rules) specifies regulations for the different zones of national parks (from Ministry of the Environment home page).

Explanation of terms in the National Park Plan

Term	Explanation	Notes
Special Protection Zone	This zone contains particularly valuable natural landscapes/wilderness and is tightly regulated.	
Special Zone Class 1	Strict regulations are used to conserve the valuable landscapes in this class.	Regulated activities require permission.
Special Zone Class 2	Farming, fishing, and forestry activities are closely regulated in this class.	
Special Zone Class 3	Conservation of the landscapes in this class is not as critical, so routine farming, fishing, and forestry activities are unregulated.	
General Zone	Located outside the Special Zone, this zone acts as a buffer to protect natural landscapes.	Regulated activities require notification.

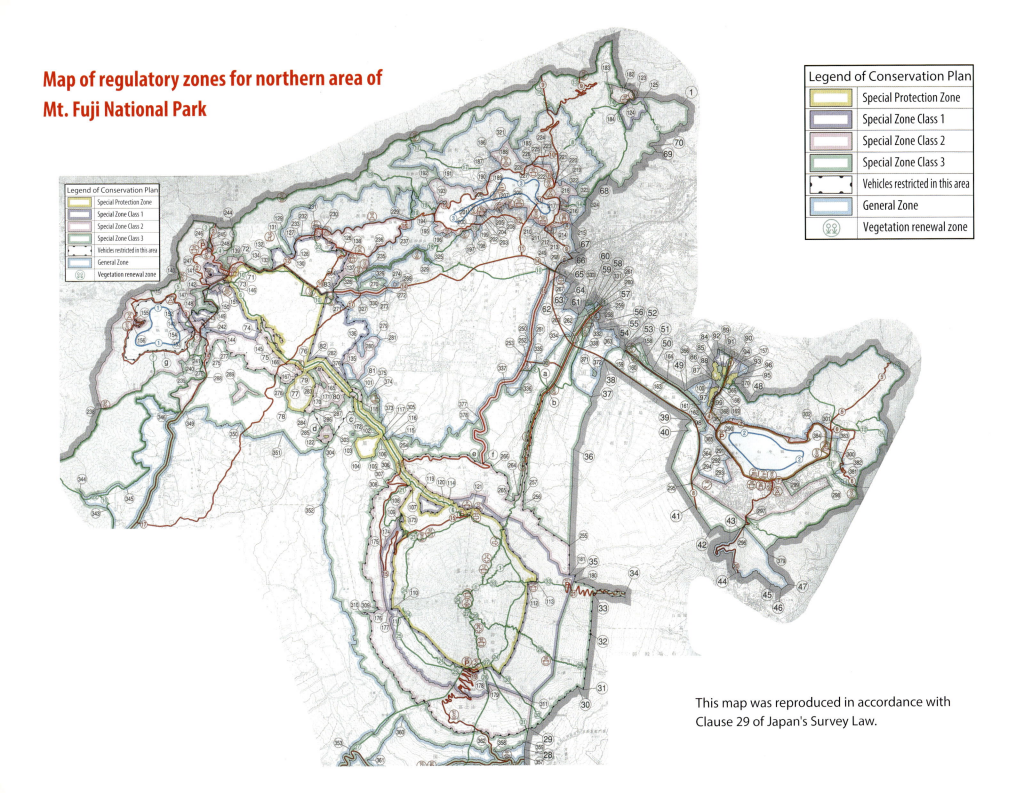

B. Nature's place in the Mt. Fuji World Cultural Heritage Site

Fuji was registered as a World Cultural Heritage Site in June 2013. With the exception of Miho-no-matsubara, the map to the right shows the 24 individual places that constitute the Mt. Fuji World Cultural Heritage Site. We can see that natural environments such as part of Mt. Fuji and the Fuji Five Lakes are included in the site, making it distinct from other World Cultural Heritage Sites in Japan.

Most of the Special Protection Zone mentioned earlier is included in the portion of Fuji considered to be part of the heritage site. That means the main climbing trails, used by 200 to 300 thousand people each year, are part of the site, as are strips of land alongside three toll roads (the Fuji Subaru Line, Fuji Azami Line, and Fujisan Sky Line) between stations 4.5 and 5. These roads are used by over one million people each year.

Thus the nature of Mt. Fuji should be conserved as part of not only a national park but also a World Cultural Heritage Site. To do that, we need to raise awareness among both domestic and foreign visitors about the precious nature found in the Special Protection Zones so that they are careful not to harm it. We hope *The Natural Features of Mount Fuji* helps visitors learn about the current status of nature around Fuji and inspires them to preserve it for future generations.

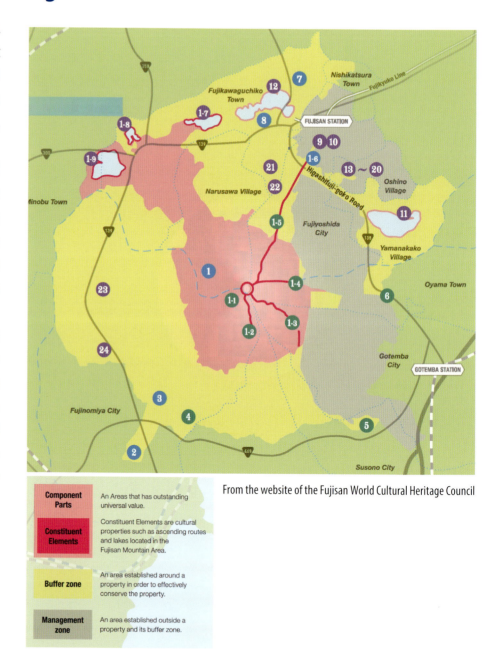

From the website of the Fujisan World Cultural Heritage Council

III. Species thought to be extinct on the northern side of Mt. Fuji

So far we have described endangered plants and animals. Here we introduce species thought to be already extinct on the northern side of Mt. Fuji.

① Ō-uragin-hyōmon (*Fabriciana nerippe*) CR(J) EX(Y&S)

The photo to the right shows a butterfly captured on July 10, 1984, the last specimen to be recorded in Yamanashi Prefecture. The most recent one collected in the Fuji region was captured by Shigeyuki Ōmori on August 6, 1965 in the Kagosaka Pass of Yamanakako Village. None have been recorded since, so this species is presumably extinct on and around Mt. Fuji.

This butterfly is considered extinct throughout most of Japan. Recently its presence has been confirmed only at Yamaguchi Prefecture's Akiyoshidai Plateau and in the area around Mt. Aso in Kyūshū. This species was listed as Critically Endangered in Yamanashi Prefecture's 2005 Red Data Book, but none of these butterflies have been spotted for over 30 years, so it appears to be extinct in all areas of Yamanashi Prefecture.

② Ō-chabane-seseri (*Polytremis pellucida*) NT(Y)

Large populations of this butterfly were widely distributed at the northern foot of Mt. Fuji until the 1960s, but the species declined from the 1970s and disappeared entirely after the 1990s. It is unclear why this species has disappeared, but today it appears to be extinct on the northern side of Mt. Fuji.

In recent years this butterfly has been recorded in the *satoyama* areas of the neighboring Misaka Mountains, and fortunately, it also continues to survive in the *satoyama* environments of mountainous areas flanking the Kōfu Basin to the east and northwest. It also lives along the Katsura River (see photo to right).

This butterfly looks like the related species **ichimonji-seseri** (Straight swift, *Parnara guttata*), but it can be distinguished by the width of the wings as well as the silver white patterns on the hindwings.

Upper side Underside

7. Mount Fuji Nature Conservation Center (NPO)

The Mount Fuji Nature Conservation Center has been engaged in three kinds of conservation projects.

I. **Research and public awareness campaigns about Mt. Fuji and nearby natural environments**
   Mt. Fuji Biodiversity Conservation Research (2012-2017)
   A. Research on the special characteristics of grassland environments at Mt. Fuji
   B. Research on the special characteristics of lava field environments at Mt. Fuji
   C. Other projects
      a. Research on changes to the distribution of invasive burr cucumber (**arechi-uri**) and its impact on endangered wetland plants
      b. Publication of *The Natural Features of Mount Fuji* in both Japanese and English

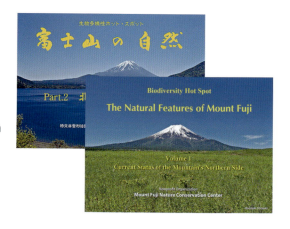

II. **Motivating elementary and high school students living in the Mt. Fuji region to conserve natural environments**
   Junior Mt. Fuji Nature Conservation Awards
   1st commendation ceremony and memorial lecture on nature photography, October 27, 2012 (Unno Kazuo presiding)
   2nd commendation ceremony and memorial lecture on nature painting, October 26, 2013 (Amano Akira presiding)
   3rd commendation ceremony and memorial lecture on insects, November 8, 2014 (Ikeda Kiyohiko presiding)
   4th commendation ceremony and memorial lecture on nature photography, November 14, 2015 (Itō Fukuo presiding)
   5th commendation ceremony and memorial lecture on nature painting, November 16, 2016 (Amano Akira presiding)
   6th commendation ceremony and memorial lecture on birds, November 18, 2017 (Higuchi Hiroyoshi presiding)

III. **Joint projects with environmental conservation groups and universities in and out of Japan**
   A. Research on the symbiotic relationship of endangered *Lycaenid* butterflies and tending ants (in cooperation with Hiroshima University)
   B. Research on the relationship between the morphology and ecology of the endangered Northeast-Asian wood white (*Leptidea amurensis*) (in cooperation with Iwate University)
   Earthwatch projects
   Lewis and Clark College Mt. Fuji Summer Program (2014 and 2017)

Lewis & Clark College Summer Program 2014

Junior Mt. Fuji Nature Conservation Awards grand prize winner, 2013

References

① Abe, H. et al. 阿部永監修・阿部永ほか(著)(2008)『日本の哺乳類改訂2版』東海大学出版会.
② Hayashi, K. et al. 林匡夫・森本桂・木元新作(編著)(1984)「原色日本甲虫図鑑(Ⅳ)」. 保育社.
③ Hayashi, Y. and Kadota, eds. 林弥栄(監修)・門田(改訂版監修)(2013) 畔上・菱山・西田(解説)「野に咲く花」. 山と渓谷社.
④ Hidaka, T., ed. 日高敏隆監修(1996)「日本動物大百科 哺乳類Ⅰ」. 平凡社.
⑤ Hiroi, T. 広井敏男(2001)「雑木林へようこそ！ 里山の自然を守る」. 新日本出版社.
⑥ Honda, K. and Y. Kato, eds. 本田計一・加藤義臣(編)(2005)「チョウの生物学」. 東京大学出版会.
⑦ Ishii, M., ed. 石井実(監修) 日本自然保護協会(編)(2005)「生態学からみた里やまの自然と保護」. 講談社.
⑧ Ishii, M., K. Higuchi and T. Shigematsu 石井実・樋口邦彦・重松敏則(1993)「里山の自然を守る」. 築地書館.
⑨ Kadota, ed. 門田(改訂版監修)・畔上(編・解説)・菱山・西田(解説)(2013)「山に咲く花」山と渓谷社.
⑩ Kanagawakenritsu Seimeinohoshi Chikyū Hakubutsukan, ed. 神奈川県立生命の星・地球博物館編(2003)「かながわの自然図鑑 3 哺乳類」. 有隣堂.
⑪ Kato, M. and A. Ebihara 加藤雅啓・海老原淳(編)(2011)「日本の固有植物」東海大学出版会.
⑫ Ministry of the Environment, ed. 環境省自然保護局野生生物課(編)(2000)「改訂日本の絶滅のおそれのあるレッドデータブック野生生物」. 自然環境研究センター．
⑬ Ito, F. and K. Maruyama; H. Kitagawa, ed. 北川尚史(監)伊藤ふくお・丸山健一郎(著)(2003)「ひっつきむしの図鑑」. トンボ出版.
⑭ Fujisabō Jimusho. Ministry of Land, Infrastructure, Transport and Tourism 国土交通省中部地方整備局富士砂防工事事務所(2002)「富士山の自然と社会」. 中部復建(株).
⑮ Miyawaki, A. and H. Sugawara 宮脇 昭・菅原久夫(1992)富士山の植物たち―典型的な垂直分布と火山植生―. 諏訪彰(編)「富士山―その自然のすべて―」: 277-294. 同文書院.
⑯ Ōkubo, E. and S. Isoda, eds. 邑田仁・鳥居塚和生(監修) 大久保栄治・磯田進(編)(2007)「富士山の植物図鑑」. 東京書籍
⑰ The Ecological Society of Japan, ed. 日本生態学会(編)(2012)「生態学入門 第2版」. 東京化学同人.
⑱ Japan Butterfly Conservation Society, ed. 日本チョウ類保全協会(編)(2012)「フィールドガイド日本のチョウ」. 誠文堂新光社.
⑲ Okino, T. 沖野外輝夫(2002)「湖沼の生態学」. 共立出版.
⑳ Satake, et al. 佐竹・大井・木村・亘理・冨成(1981)「日本の野生植物Ⅲ」. 平凡社.
㉑ Satake, et al. 佐竹・大井・木村・亘理・冨成(1982)「日本の野生植物Ⅰ」. 平凡社.
㉒ Satake, et al. 佐竹・大井・木村・亘理・冨成(1982)「日本の野生植物Ⅱ」. 平凡社.
㉓ Shigamatsu T. and JCVN, eds. 重松敏則＋JCVN(編)(2010)「よみがえれ 里山・里地・里海」. 築地書館.
㉔ Shimada, M. et al. 嶋田正和・山村則男・粕谷英一・伊藤嘉昭(2005)「動物生態学 新版」. 海游舎.
㉕ Shirouzu, T. 白水 隆(2006)「日本産蝶類標準図鑑」. 学研.
㉖ Shinzato, T. and M. Takeda 新里達也・武田雅志(2009)多摩川流域におけるヒメビロウドカミキリ個体群の分布と保全. 特定非営利活動法人野生生物調査協会.
㉗ Takano, S. 高野伸二(2008)「フィールドガイド日本の野鳥」(増補改訂版). 日本野鳥の会.
㉘ Watanabe, M. 渡辺 守(2007)「昆虫の保全生態学」. 東京大学出版会.
㉙ Uesugi, Y., ed. 上杉 陽(編)「地学見学案内書 富士山」. 日本地質学会関東支部.
㉚ Yamanashiken Shinrinkankyōbu Midori-shizenka, ed. 山梨県森林環境部みどり自然課(編)(2005)「2005 山梨県レッドデータブック」. 山梨県.
㉛ Inoue, H. et al.; Yata, O., ed. 矢田脩(監)井上寛ほか(著)(2007)「新訂原色昆虫大図鑑 第1巻(蝶蛾篇)」. 北隆館.
㉜ Watanabe, Michihito and Yasuo Hagiwara (2009) A newly observed form of symbiotic relationship between Reverdin's blue *Lycaeides argyrognomon praeterinsularis* (Varty), (Lycaenidae) and *Camponotus japonicus* Mayr (Formicidae). Journal of Research on the Lepidoptera 41: 70-75

INDEX

Geological and geographycal words

Aokigahara	53,56
Driftwood mold	55
First Hōei crater	54
Fugaku Fūketsu Cave	55
Fuji Five Lakes	62,63,65,66,67,68,76
Fuji Fūketsu Cave	54,56
Fujinomiya Ōsawa Valley	22,54
Funatsu Tainai	55
Hōei eruption	54
Horizontal mold	55
Katsura River	62,63,77
Kenmarubi 1	53,55
Kenmarubi 2	53,55
Kōfu Basin	70,71,77
Lake Kawaguchiko	72,73
Lake Yamanakako	62,69
Lava cave	54,58,61
Lava flow	53,54,55,56
Lava tree mold	54,55,61
Lava tube	56,57
Lava tunnel	56
Leaning mold	55
Loam layer	63
Motosu Fūketsu Cave No. 1	56
Motosu Fūketsu Cave No. 2	56
Mt. Nagao	56
Narusawa Hyōketsu Cave	56
Neogene period	63
New Fuji	53,54
Old Fuji mudflows	62,63
Oshino Hakkai	62
Parasitic cone	56
Pyroclastic flow	63
Pyroclastic material	54,62,63
Ropey lava	56,57
Saiko Bat Cave	54,56,57
Stone pillar mold	55
Takamarubi	53
Well mold	55
Yoshida Tainai	55

Plants

<Japanese Name>

Ao-tenma	39
Arechi-uri	72,78
Arinotōgusa	17
Asa	27
Baikamo	69
Bāsobu	16
Benibana-yama-shakuyaku	41,43
Enoki	42
Ezo-enoki	42
Fugaku-suzumushisō	38
Funabarasō	18
Hanahatazao	19
Himetsuruninjin	16
Hinanokinchaku	17
Igusa	66
Jakōchidori	16
Jigabachisō	38
Jisobu	16
Jōrōsuge	65
Kaijindō	18
Kawajisha	67
Kayatsuri-suge	68
Kikyō	13
Kitsunenobotan	68
Ko-atsumorisō	41
Ko-kitsunenobotan	68
Ko-matsukasa-susuki	67
Konara	36
Kōrinka	14
Matsukasa-susuki	67
Mikuri	69
Mizuchidori	16
Murasaki	19
Murasaki-senburi	15
Okinagusa	13
Oniku	74
Oninoyagara	39
Sakane-ran	39
Senburi	14
Suge	68
Sujinumahari'i	66
Suzumushisō	38
Suzusaiko	15
Takonoashi	66
Tenma	39
Yama-shakuyaku	40,41,43

<English Name>

Balloon flower	5,13
Burr cucumber	72,78
Bush clover	33
Hackberry	42
Hemp	27
Japanese beech	8
Japanese honeysuckle	25
Japanese oak	9
Japanese pampas grass	9
Killifish	5
Mistletoe	49
Mongolian oak	8
Oak	36
Peony	40
Pinecone	48,67
Smartweed	33
Thistle	27
Water speedwell	67

<Binomial(Latin) Name>

Acalolepta degenera	27
Ajuga ciliata	18
Artemisia japonica	27
Asclepiadoideae	18
Boschniakia rossica	74
Campanulacea	16
Carex	33
Carex bohemica	68
Carex capricornis	65
Carex lanceolata	22,23
Celtis jessoensis	42
Codonopsis ussuriensis	16
Cypripedium debile	41
Dianthus superbus	23

Dontostemon dentatus	19
Eleocharis equisetiformis	66
Elymus tsukushiensis	22
Gastrodia elata	39
Haloragaceae	17
Haloragis micrantha	17
Liparis fujisanensis	38
Liparis krameri	38
Liparis makinoana	38
Lithospermum erythrorhizon	19
Miscanthus	34
Miscanthus sinensis	9,22,23,34
Neottia nidus-avis	39
Orchidaceae	41
Orchis	38,39
Paeonia japonica	40
Paeonia obovata	41
Patrinia scabiosifolia	25
Penthorum chinense	66
Platanthera hologlottis	16
Platycodon grandiflorus	13
Poaceae	33,35
Polygala tatarinowii	17
Polygalaceae	17
Polygonaceae	33
Potentilla fragarioides	22
Potentilla freyniana	22
Pulsatilla cernua	13
Ranunculus chinensis	68
Ranunculus nipponicus	69
Rhamnus davurica	23
Sanguisorba officinalis	21,24
Scirpus mitsukurianus	67
Sicyos angulatus	72
Sparganium erectum	69
Spodiopogon sibiricus	22
Swertia japonica	14
Swertia pseudochinensis	15
Tephroseris flammea	14
Veronica anagallis-aquatica	67
Vicia amoena	24,25
Vincetoxicum atratum	18
Vincetoxicum pycnostelma	15

Insects

\<Japanese Name\>

Aka-seseri	22
Asa-kamikiri	27
Asama-shijimi	20
Benimon-madara	25
Chamadara-seseri	22
Futasuji-katabiro-hanakamikiri	43
Gaganbo	60
Gin'ichimonji-seseri	22
Goma-shijimi	21
Herigurochabane-seseri	22
Himebirōdo-kamikiri	27
Hime-shijimi	20
Himeshirochō	24
Hoshichabane-seseri	22
Hyōmonchō	24
Ichimonji-seseri	77
Jigabachi	38
Kimadara-modoki	23
Kuro-shijimi	21
Madara-kamadouma	60
Maemon-ō-namishaku	61
Miyama-shijimi	20
Ō-chabane-seseri	77
Ōmurasaki	42
Ō-uragin-hyōmon	77
Sujigurochabane-seseri	22
Sukiba-hōjaku	25
Suzumushi	38
Uraginsuji-hyōmon	25
Yama-kichō	23
Yokozuna-sashigame	73

\<English Name\>

Bell cricket	38
Brimstone	23
Crane flies	60
Eastern silverstripe	25
Great purple emperor	42
Japanese scrub hopper	22
Long-horned beetles	27,43
Maculatus skipper	22
Marbled fritillary	24
Northeast-Asian wood white	24,78
Oriental checkered darter	22
Pseudo-labyrinth	23
Reverdin's blue	20,26
Silver-lined skipper	22
Silver-studded blue	20
Straight swift	77

\<Binonial(Latin) Name\>

Acalolepta degenera	27
Aeromachus inachus	22
Agriosphodrus dohrni	73
Argyronome laodice	25
Brachyta bifasciata	43
Brenthis daphne	24
Camponotus japonicus	20,21,26
Camponotus obscuripes	20
Fabriciana nerippe	77
Formica japonica	20
Formica yessensis	20
Geometrid	61
Gonepteryx rhamni	23
Hemaris radians	25
Hesperia florinda	22
Kirinia fentoni	23
Lasius japonicus	20
Leptalina unicolor	22
Leptidea amurensis	24,78
Lycaenid	20,78
Maculinea teleius	21
Myrmica kotokui	21
Niphanda fusca	21
Parnara guttata	77
Plebejus subsolana	20
Plebejus argus	20
Plebejus arygyrognomon	20
Polytremis pellucida	77
Pristomyrmex punctatus	20
Pyrgus maculatus	22
Sasakia charonda	42
Thyestilla gebleri	27
Thymelicus leoninus	22
Thymelicus sylvaticus	22
Triphosa sericata	61
Zygaena niphona	24,25

Birds

\<Japanese Name\>

Akamozu	29
Buppōsō	47
Fukurō	45
Hagi-mashiko	33,44
Haiiro-chūhi	32
Hai-taka	44
Hayabusa	31
Hi-renjaku	49
Isuka	48
Keashi-nosuri	32
Ki-renjaku	49
Ko-chōgenbō	31
Ko-mimizuku	32
Kuroji	48
Misago	31
Nojiko	47
Nosuri	32
Ōjishigi	28
Ōkanada-gan	73
Ōkonohazuku	48
Ō-mashiko	33
Ō-taka	44
Renjaku	49
Sankōchō	46
Sanshōkui	47
Shijūkara-gan	73
Toratsugumi	46
Yotaka	30

\<English Name\>

Ashy minivet	47
Asian rosy finch	33,44
Bohemian waxwing	49
Broad-billed roller	47
Brown shrike	29
Bunting	47
Canada goose	73
Common crossbill	48
Crow	31,32,44
Dove	31,44
Flycatcher	46
Grey bunting	48
Grey nightjar	30
Grey starling	46,47
Hawk	44
Hen harrier	32
Japanese paradise flycatcher	46
Japanese waxwing	49
Japanese yellow bunting	47
Kite	31,32
Latham's snipe	28
Meadow bunting	48
Merlin	31
Northern goshawk	44
Osprey	31
Owl	45
Peregrine falcon	31
Pigeon	44,48
Rosefinch	33
Rough-legged buzzard	32
Sandpiper	28
Short-eared owl	32
Shrike	29,49
Snipe	28
Sparrow	33,46
Sparrowhawk	44
Sunda scops owl	48
Thrush	46
Ural owl	45
Wagtail	47
Waxwings	49
White's thrush	46

\<Binonial(Latin) Name\>

Accipiter gentilis	44
Accipiter nisus	44
Asio flammeus	32
Bombycilla garrulus	49
Bombycilla japonica	49
Branta canadensis	73
Buteo lagopus	32
Caprimulgus indicus	30
Carpodacus roseus	33
Circus cyaneus	32
Emberiza sulphurata	47
Emberiza variabilis	48
Eurystomus orientalis	47
Falco columbarius	31
Falco peregrinus	31
Gallinago hardwickii	28
Lanius cristatus	29
Leucosticte arctoa	33
Loxia curvirostra	48
Otus lempiji	48
Pandion haliaetus	31
Pericrocotus divaricatus	47
Strix uralensis	45
Terpsiphone atrocaudata	46
Zoothera dauma	46

Mammals

\<Japanese Name\>

Araiguma	73
Kaya-nezumi	34
Kikugashira-kōmori	57,58
Ko-kikugashira-kōmori	59
Momojiro-kōmori	58,59
Tengu-kōmori	59
Usagi-kōmori	58,61
Yamane	50,51

\<English Name\>

Brown long-eared bat	58
Common raccoon	73
Greater horseshoe bat	58
Harvest mouse	34
Japanese dormouse	50
Japanese grass vole	45
Japanese large-footed bat	59
Little Japanese horseshoe bat	59
Mice	45,48
Mouse	34
Rabbit	44
Tube-nosed bat	59

\<Binonial(Latin) Name\>

Glirulus japonicus	50
Micromys minutus	34
Murina hilgendorfi	59
Myotis macrodactylus	59
Plecotus auritus	58
Procyon lotor	73
Rhinolophus cornutus	59
Rhinolophus ferrumequinum	58

Other Animals

\<Japanese Name\>

Mamushi	71
Moriao-gaeru	52
Shima-hebi	71
Tōkyō-daruma-gaeru	70
Tonosama-gaeru	70

\<English Name\>

Black-spotted pond frog	5,70
Forest green tree frog	52
Japanese striped snake	71

\<Binonial(Latin) Name\>

Elaphe quadrivirgata	71
Gloydius blomhoffii	71
Pelophylax nigromaculatus	70
Rhacophorus arboreus	52

Other words

APG (Angiosperm Phylogeny Group) plant classification system	15,18
"Araiguma rasukaru"	73
Aposematic coloration	25
Bōga kōshin	36,37
Earthwatch projects	78
Facultative symbiosis	20
Feigning injury	30
Firebreak zone	11
"Foster" type symbiosis	21
Fuji Azami Line	76
Fuji Subaru Line	76
Fuji-Hakone-Izu National Park	74
Fujisan Sky Line	76
Fujisan World Cultural Heritage Council	76
Ground nest	35
Half-underground nest	35
Horizontal biodiversity	53
Ice "bamboo shoot"	60
Invasive Alien Species	72,73
Jakō	16
Junior Mt. Fuji Nature Conservation Awards	78
Kagosaka Pass	77
Kayaba	4
Lewis and Clark College	78